Applications
of the
Theory of Plasticity
in
Soil Mechanics

WILEY SERIES IN GEOTECHNICAL ENGINEERING

Consulting Editors

T. W. Lambe
R. V. Whitman

Department of Civil Engineering
Massachusetts Institute of Technology

**Applications of the Theory of Plasticity
in Soil Mechanics**

J. Salençon

Ecole Polytechnique
Ecole Nationale des Ponts et Chaussées, Paris

Applications of the Theory of Plasticity in Soil Mechanics

J. Salençon
Ecole Polytechnique
Ecole Nationale des Ponts et Chaussées, Paris

Translation by
R. W. Lewis and H. Virlogeux
Revised by the author

A Wiley–Interscience Publication

JOHN WILEY & SONS
Chichester . New York . Brisbane . Toronto

First published under J. Salençon *Théorie de la plasticité pour les applications à la mécanique des sols* by Editions Eyrolles-Paris.

© 1974 Eyrolles

English translation Copyright © 1977, by John Wiley & Sons, Ltd.

All rights reserved.

No part of this book may be reproduced by any means, nor transmitted, nor translated into a machine language without the written permission of the publisher.

Library of Congress Cataloging in Publication Data:
Salençon, Jean.
    Applications of the theory of plasticity in soil mechanics.

    'A Wiley–Interscience publication.'
    Translation of Théorie de la plasticité pour les applications à la mécanique des sols.
    Include; bibliographies and index.
    1. Soil mechanics. 2. Plasticity. I. Title.
TA710.S2413   624'.1513   77-611

ISBN 0 471 74984 2

Printed in Great Britain by Page Bros (Norwich) Ltd, Norwich.

To
Professor Jean Mandel

Contents

Foreword xi

Chapter I General considerations on the plastic behaviour of materials
1 Yield criterion 1
2 Form of the yield criteria 3
3 Plastic deformation 8
4 The origin of the plastic deformation 9
5 Flow rule 9
6 Principle of maximum plastic work 12
7 Consequences of the constitutive law 13
8 Validity of the principle of maximum plastic work for the case of soils 14
9 Final remarks 15
10 Application of the plasticity model to soil behaviour 16
References 16

Chapter II A short survey of the problems of elasto-plasticity
1 Introduction 19
2 Statement of the problem. Method of solution 19
3 Behaviour of a system subject to a loading process 20
4 Importance of the geometry changes 22
5 Examples 24
References 26

Chapter III The problems of uncontained plastic flow and an investigation of the 'rigid-plastic model'
1 Definition of the rigid-plastic material 28
2 Statement of the problem for a rigid-plastic system 28
3 The rigid-plastic pattern and the determination of limit loading . 29
4 Governing equations 31
5 Boundary conditions 32
6 Theorem of uniqueness of the stress-field 35
7 Remarks 38
References 38

Chapter III **Appendix**   40
  General definition of the loading parameters for a system 40
    1 Possible loadings 40
    2 Loading process depending on a finite number of parameters . 41
    3 Case of a system with friction conditions 43
    4 Example 43

Chapter IV **Problems of uncontained plastic flow in plane strain**
    1 General 45
    2 Expression of the yield criterion 45
    3 A remark on the case of a non-standard material 47
    4 Equations for the stresses 47

  A The stress problem 48
    5 Transformation of the equations 48
    6 Characteristic lines 50
    7 Relations along the characteristic lines 50
    8 Computation of the solution 51
    9 Transformation of equations (24, 25) 53
    10 Geometry of the characteristic network 54
    11 Matching of solutions 56

  B The velocity problem 56
    12 Flow rule 56
    13 Characteristics—Relations for the velocities along the characteristics 58
    14 Examples 59
    15 Discontinuity of the velocity 61
    16 Discontinuity of the stress-field 62

  C Study of an example 62
    17 The problem 62
    18 Construction of the solution 63
    19 Calculation of the velocities 63
    20 A particular case 64
    21 Remarks on the solution 65
    22 The case of a Coulomb material 65

  D Particular uses in the study of a Coulomb material 67
    23 Method of superposition—Theorem of the corresponding states 67
    24 An example study of a cohesionless soil with self-weight . . 70
    References 74

Chapter IV Appendixes

A Problems of uncontained plastic flow in plane strain for isotropic
   non-homogeneous material 77
   1 General 77
   2 The problem for the stresses—General case 77
   3 The case of a Tresca material 81
   4 The case of a Coulomb material. 81
   5 The velocity problem. 82
   6 The case of some non-standard materials 86
   7 Discontinuity of the stress-field 89
   References. 95

B Problems of uncontained plastic flow with axial symmetry for
materials with a yield criterion of the 'intrinsic curve' type 96
   1 General 96
   2 The stress problem 97
   3 The velocity problem 99
   4 Weak solutions 102
   References. 102

Chapter V The theory of limit analysis (For applications to soil mechanics)

   1 Presentation 104
   2 Admissible fields. Dissipation 105
   3 Static approach 108
   4 Kinematic approach 112
   5 Remarks on the results of the theory of limit analysis . . . 113
   6 Minimum principles 114
   7 Limit analysis in the study of plane strain problems of uncontained
     plastic flow 115
   8 Interpretation of the solutions for the case of a Coulomb material 119
   9 Other applications of limit analysis in soil mechanics 122
   References. 126

Chapter V Appendixes

A An alternative presentation of the theory of limit analysis. Extension
to some non-standard materials 129
   1 Introduction 129
   2 The case of standard material 129
   3 Case of non-standard systems 141
   4 Friction conditions at the interfaces 146
   References. 150

B Bonneau's theorem 150
   1 The problem 151
   2 Origin of the problem—The case of incomplete solutions . . 151
   3 Lemma 152
   4 Statement of the differential relations necessarily verified along (C) 153
   5 Remarks 156
      References 156

Index 157

Foreword

The purpose of this work is not the training of specialists in Plasticity; it aims at helping the reader to a better understanding of Soil Mechanics so far as it appeals to the theory of plasticity, so as to be able, if necessary, to appreciate the meaning and validity of available computational methods.

We have been trying to gather from the theory of Plasticity whatever can prove useful in Soil Mechanics, in the present state of the art; such matters have been exposed with some generality, whilst others, such as the phenomenon of plastic adaptation and the dynamical problems, have not been evoked.

There are five main headings.

The first chapter is devoted to the presentation of the classical model of plastic behaviour, and to the possibilities of the application of this model in the case of soils; in particular, the problem of the flow rule is studied in detail, in the light of recent research.

In the second chapter, the problems of elasto-plasticity are dealt with briefly. These are rather difficult, and, with the exception of certain cases in which simplifications appeared, they have received scant application to the matters which interest us. However, it is essential to realize that the elasto-plastic problem is the basic problem as soon as we introduce plasticity (without viscosity) into the material behaviour; and that, as a consequence, it is essential to see for which conditions, under which hypotheses, and for which types of problems, we can return to the rigid-plastic scheme.

This question, in its generality, constitutes the object of the third chapter. The various ways of defining and considering the rigid-plastic material are examined, and the method of stating limit equilibrium problems is studied.

The fourth chapter is devoted to a detailed study of the theory of plane limit equilibrium, the uses of which are very numerous in theoretical Soil Mechanics.

Finally, the theory of limit analysis is dealt with in Chapter V. Firstly the case of a material with an associated flow rule is examined, and we determine accurately the range of application of the results obtained, particularly as regards the significance of the rigid-plastic scheme. A clear understanding of the calculation processes used for the rigid-plastic material can be attained only through the results of the theory of limit analysis. The case of a material with a non-associated flow rule is also considered, the deficiencies in this case being shown, together with the positive points that can be used as foundations.

Chapters III, IV, and V are followed by one or several appendixes in which are given developments which would have made the text too heavy.

In each chapter and appendix numbers placed in square brackets in the text refer to a list of references. Though not attempting to present an exhaustive bibliography, these lists indicate interesting works the reader can refer to.

Finally, it should be noted that, except for the examples that are given for a didactic purpose, no solutions of more or less classical problems of Plasticity linked to Soil Mechanics are to be found in this work. Effectively, our purpose is not to replace the well-known books of Sokolovski, Hill, etc., but rather to prepare the reader for their possible utilization.

I want, finally, to pay homage to Professor Jean Mandel, who was my master, and whose imprint this book bears.

I would like to thank Professor Legrand who holds the Chair of Soil Mechanics at the Ecole des Ponts at Chaussées, and who invited me to deliver the course of lectures on which this book is based. I would also like to thank Professors Habib and Radenkovic from whom I have learnt a lot in both theory and the practice of the subject in many friendly discussions.

Mrs. Hélène Virlogeux and Dr. R. W. Lewis undertook the task of translating this book into English and I would like to thank them for their patience, as I feel the readers will be well satisfied with the quality of their work.

J.S.

NOTATIONS: following the ordinary practice in Continuum Mechanics, and differing from that of Soil Mechanics, the stresses are denoted positively for tension.

CHAPTER I

General considerations on the plastic behaviour of materials

This chapter defines the plastic behaviour of materials and gives the corresponding constitutive laws. The viscosity effects of the materials are ignored and hence the behaviour is independent of physical time. The subjects dealt with are therefore elasticity and plasticity and not visco-elasticity or visco-plasticity.

Finally, a discussion on the application of this behaviour pattern is outlined for the case of soils.

1. Yield Criterion

Plasticity is characterized by the existence of a *yield point* beyond which *permanent strains* appear.

Consider for example a simple tension test in which the stresses and strains are assumed to be homogeneous. Figure I.1a depicts the stress–strain relationship.

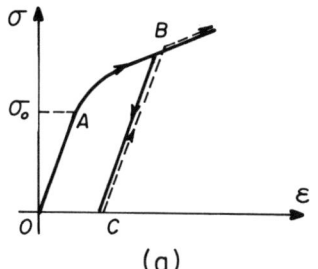
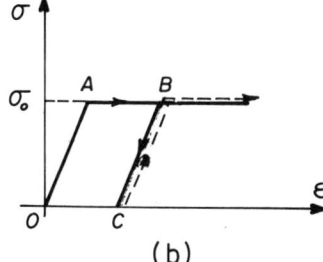

Figure I.1

Along the path OA the behaviour is elastic; i.e. the path is reversible. However, once point A has been passed (i.e. $\sigma > \sigma_0$), the path is no longer reversible. If, for instance, the sample is unloaded after reaching point B and the stress σ falls to zero, then the unloading path is given by BC, which is parallel to OA (provided the linear elastic properties of the material are not altered by the plastic deformation). After unloading there remains a strain ε^p, represented by

OC, and, termed the permanent strain. The stress σ_0 is defined as the original yield point. If the test specimen is again loaded the path is reversible along CB, becoming irreversible when $\sigma > \sigma_B$. In this case the stress σ_B is defined as the current yield point.

Figure I.1a represents the case in which σ_B is a function of the permanent strain ε^P and thus illustrates *work-hardening*. In Figure I.1b the stress σ_B is a constant and the material is said to be *perfectly plastic*.

It must be emphasized that the onset of plasticity is not indicated by the non-linearity of the stress–strain curve beyond point A but by the *irreversibility* of the path beyond this point.

More generally, it has been proved that the concept of the yield point in the uniaxial case may be replaced by a *yield criterion* for a small (macroscopic) element of material subject to any action characterized by a tensor of applied stresses $\boldsymbol{\sigma}$. f is a scalar function of the state of stress of the material, such that $f(\boldsymbol{\sigma}) < 0$ corresponds to the elastic range of the material and $f(\boldsymbol{\sigma}) = 0$ corresponds to the appearance of the irreversible deformations.[1]

Usually, it is the equality

$$f(\boldsymbol{\sigma}) = 0 \tag{1}$$

which is termed the *yield criterion*. The function f is often called a *yield function* and the surface $f(\boldsymbol{\sigma}) = 0$ in the stress-space $\{\boldsymbol{\sigma}\}$ is the *yield* or *loading surface*[2] of the material.

With a *perfectly plastic* material, the yield function does not vary and the yield surface is fixed, with plastic strains occurring only if $\boldsymbol{\sigma}$ is on this surface and stays on it.

With a work-hardening material, the yield function varies as the permanent deformation continues and discrimination must be made between the original and current yield surfaces. Additional plastic strains appear only if $\boldsymbol{\sigma}$ is situated

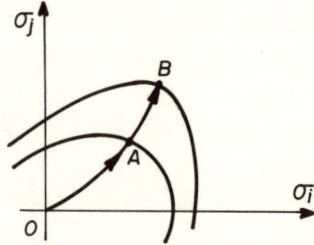

Figure I.2

---

[1] Strictly speaking, this classical representation of the yield criterion as a function of the Eulerian stress-tensor is not correct (it does not respect the principle of objectivity), with the exception of the particular case of an isotropic material (see [33], p. 716). However, this form will be used since only the case of isotropic materials will be studied later.

[2] Function f being physically determined only on symmetrical tensors, since $\boldsymbol{\sigma}$ is symmetrical ($\sigma_{ij} = \sigma_{ji}$ or $\boldsymbol{\sigma}^T = \boldsymbol{\sigma}$), its mathematical expression, which depends on nine $\boldsymbol{\sigma}$ components, is somewhat arbitrary. The mathematical expression chosen is the one symmetrical with respect to the components σ_{ij} and σ_{ji}; further on, it will be denoted by f.

on the yield surface and moves outward. The yield surface is now extended by σ, as represented graphically in Figure I.2.

In order to take the work-hardening effect into account the yield criterion will now be written as

$$f(\sigma, E) = 0 \qquad (2)$$

where E stands for all the work-hardening parameters. These are defined as the supplementary parameters which, along with the actual state of stress, make it possible to determine the plastic 'behaviour' of the material. In most instances such parameters are functions of the stress–strain history.

Various theories have been proposed in order to explicate a form for the hardening parameters. Among others are the hypothesis of a unique parameter corresponding to the work expended in the plastic deformation (see [21], and Mandel's theory [32]).

2. Form of the Yield Criteria

2.1 Convexity

It will be seen in Section 5 concerning the relation between stress and strain that a hypothesis commonly accepted for metals and some other materials, the principle of maximum plastic work, implies that the yield surface is convex in the space {σ}. It follows that f can then be chosen as a convex function of σ. This means, from a mathematical viewpoint, that if σ^1 and σ^2 verify

$$f(\sigma^1) \leqslant 0 \quad \text{and} \quad f(\sigma^2) \leqslant 0$$

then for $\forall \lambda \in (0, 1)$,

$$f[\lambda\sigma^1 + (1-\lambda)\sigma^2] \leqslant 0$$

Then the yield surface is a convex surface in the stress-space {σ}. The author believes that the convexity of the yield function[1] can be considered as a generally valid feature for normal materials, including those for which the principle of maximum plastic work can be neither proved nor even admitted.

In fact, it is sufficient that each of the elementary mechanisms necessary for plastic flow in a small macroscopic element corresponds to a convex condition for σ. The elastic range of the element is then determined by the intersection of the plastic ranges of each mechanism and is therefore a convex domain in the space {σ}.

2.2 Dealing with material symmetries

It is possible to be more explicit concerning the form of the yield criteria for the case of material symmetries. In fact, the yield criterion must deal with these symmetries.

---

[1] Or 'loading function'.

Thus, in the case of a material which was originally isotropic, f in equation (1) depends only on the *invariants of* the tensor **σ** [56], or may be expressed in the form of a *symmetrical function of the principal stresses* $\sigma_1, \sigma_2, \sigma_3$.

If isotropy is preserved throughout the work-hardening[1] period the function f in equation (2) also depends only on the invariants of the tensor **σ**, but its expression varies with the load. This assumption of isotropic work-hardening is a theoretical conception that may be admitted provided the deformations are not excessive. The most important yield criteria used for isotropic materials will be studied later.

2.3 Von Mises' criterion

For ductile materials experience has shown that if the tensor **σ** is decomposed to separate its deviator **s** in the form

$$\sigma_{ij} = s_{ij} + \frac{\sigma_{ii}}{3}\delta_{ij} \qquad (3)$$

(δ_{ij} being components of Kronecker's unit tensor, **1**), f depends only on **s**. The material plastic behaviour requires no modification on adding any isotropic state of stress.

For an isotropic, ductile material f depends only on the invariants of **s**:

$$J_1 = 0, \quad J_2 = \tfrac{1}{2}S_{ij}S_{ij}, \quad J_3 = \tfrac{1}{3}s_{ij}s_{jk}s_{ki}.^2$$

The simplest criterion of this type may be stated as

$$J_2 - k^2 = 0 \qquad (4)$$

where k^2 is a constant. This is known as the Von Mises criterion. It can easily be checked that in the stress-space $\sigma_1, \sigma_2, \sigma_3$ the yield surface is a cylinder of revolution about the vector (1, 1, 1). The yield point in pure shear is given by k. The limit in pure tension is $k\sqrt{3}$.

2.4 Tresca's criterion

Tresca's criterion also is valid for isotropic, ductile materials. It may be stated as a function of the principal stresses, ordered according to $\sigma_1 \geqslant \sigma_{11} \geqslant \sigma_{111}$.

The criterion is

$$\sigma_1 - \sigma_{111} = 2k \qquad (5)$$

where the intermediate principal stress σ_{11} plays no part.

---

[1] This is, in particular, the case for 'Taylor's isotropic work-hardening', which depends on one scalar parameter.

[2] Unless the contrary is specified, the summation convention for repeated subscripts is used.

With respect to the unordered principal stresses, $\sigma_1, \sigma_2, \sigma_3$, the yield function assumes the symmetrical form

$$f(\boldsymbol{\sigma}) = \sup_{\substack{i=1,2,3 \\ u=1,2,3}} \{\sigma_i - \sigma_j - 2k\} \quad (6)$$

It will be verified that in the space $\sigma_1, \sigma_2, \sigma_3$ the yield surface is a hexagonal prism parallel to the axis $(1, 1, 1)$. The yield point in pure shear is given by k. The limit in pure tension is $2k$.

According to the results of Section 2.2, the Tresca yield-function (equation (6)) can also be expressed by means of the invariants of $\boldsymbol{\sigma}$ (and, in this case, the invariants of $\mathbf{s}$). In the present case this raises some problems since a closed form in J_2, J_3 equivalent to (6) does not exist,[1] and presents no practical interest.

2.5 Intrinsic curve (Mohr, Caquot)

Mohr, followed by Caquot, proposed a generalization of Tresca's criterion for an isotropic material:

$$\sigma_1 - \sigma_3 = g(\sigma_1 + \sigma_3) \quad (7)$$

the principal stresses $\sigma_1, \sigma_2, \sigma_3$ being ordered according to $\sigma_1 \geqslant \sigma_2 \geqslant \sigma_3$. (As no ambiguity can be expected the distinction between Arabic and Roman indices[2] is no longer maintained.) The function g must be determined experimentally and is found to have the following properties:

$g \searrow$ if $(\sigma_1 + \sigma_3) \nearrow$; it becomes zero for a positive value of $(\sigma_1 + \sigma_3)$ when $(\sigma_1 + \sigma_3)/2$ is the isotropic tensile strength; and when $(\sigma_1 + \sigma_3) \searrow -\infty$, g tends to a limit so that Tresca's criterion is found asymptotically (see [33]).

The characteristic of this type of yield criterion is that the intermediate principal stress has no influence.

Equation (7) gives a relationship between the radius $(\sigma_1 - \sigma_3)/2$ and the abscissa $(\sigma_1 + \sigma_3)/2$ of the centre of a Mohr's circle in accordance with the state of stresses at the point under consideration. This verifies that Mohr's circles representing all the limit states have an envelope, termed the intrinsic curve.

Adopting Hill's notation gives the expression

$$-p = (\sigma_1 + \sigma_3)/2$$

For the stress representation given by Mohr, Figure I.3 shows Mohr's circles corresponding to various states of limit equilibrium and the intrinsic curve enveloping these circles.

$$|\tau| = h(\sigma) \quad (8)$$

is the equation of this curve, where σ and τ are the normal and tangential components of the stress on a plane.

---

[1] The closed polynomial expression in J_2, J_3 given in [44] is not correct.
[2] The expression for $f(\boldsymbol{\sigma})$ as a function of the unordered principal stresses is analogous to (6).

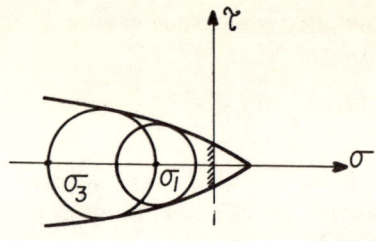

Figure I.3

For a material having a yield criterion given by equation (7) the following condition applies.

In order that the stress state at a point M does not violate the yield criterion, i.e. it satisfies $f(\boldsymbol{\sigma}) \leqslant 0$, it is necessary and sufficient that Mohr's circle does not intersect the intrinsic curve; thus,

$$|\tau| \leqslant h(\sigma) \qquad (9)$$

is valid for all the planes through point M.

If, and only if, the yield criterion is reached, there exist two planes on which $|\tau| = h(\sigma)$.[1] These planes are symmetrical to each other with respect to the planes of the major and minor principal stresses.

2.6 Coulomb's criterion

Coulomb's yield criterion is very often used for soils and is of the 'intrinsic curve' type in which the intrinsic curve is constituted by two symmetrical straight lines inclined at an angle ϕ towards the axis of normal stresses.

With the usual notation, $H = C \cot \phi$ and equation (7) becomes

$$(\sigma_1 - \sigma_3) = -\sin \phi (\sigma_1 + \sigma_3 - 2H) \qquad (10)$$

and (9) becomes

$$|\tau| \leqslant C - \sigma \tan \phi.\text{[2]} \qquad (11)$$

In the stress space $(\sigma_1, \sigma_2, \sigma_3)$ the yield surface is a hexagonal pyramid which admits the axis $(1, 1, 1)$ as a ternary axis of symmetry and the bisector planes as planes of symmetry (a consequence of isotropy). The cross section has the form of a hexagon. Complementary data in this respect will be found in classical works (for example [30]).

---

[1] It will be assumed that the relation $|\tau| = h(\sigma)$ is real; i.e. the family of circles has a real envelope. Equation (8) represents two arcs symmetrical about 0σ. The problem of the shape of the intrinsic curve has been discussed ('flat' or 'angular' summit); according to the author's opinion, there is no real difficulty. One must not forget, moreover, that most often, the physical phenomenon corresponding to the vicinity of the summit of the intrinsic curve has a different nature: brittle fracture occurs without plastic yielding.

[2] The somewhat unusual minus sign is due to the adopted sign convention. Hill's convention and other common notations are chosen to make further reading easier.

2.7. Drucker–Prager criterion

Drucker and Prager [11] proposed a criterion for soils that is related to Mises criterion just as Coulomb's criterion is related to that of Tresca. The yield surface is a right circular cones with axis (1, 1, 1). It is regular, a fact which can prove advantageous for some calculations.

The loading function for this criterion is written as a function of the invariants of the stress tensor and of the deviator. If

$$I_1 = \sigma_{ii}$$

it assumes the form

$$f = \alpha I_1 + J_2^{1/2} - k$$

in which α and k are positive constants.

Other authors also proposed a yield criterion for which the yield surface forms a regular hexagonal pyramid with an axis (1, 1, 1).

2.8 Plasticity and fracture

As observed in Section 2.5 it is useful in this study of yield criteria to take fracture into consideration. It is not intended to deal at great length with this subject, which is now in full development.

Usually a fracture—i.e. the separation of a solid into several pieces—can appear in either of two ways.

(1). Without previous plastic deformation it is termed *brittle fracture*, which occurs by fracture along a surface normal to the direction of the greatest tensile stress.

(2). After previous plastic deformation it is termed *ductile rupture*.

Considering the phenomenon in a simplified manner the elastic range of the material can be conceived as the interaction of two ranges which are probably convex. One of these corresponds to the limit with regards to brittle fracture, and the other to the appearance of plastic deformations.

The yield surface is the boundary of this range and it appears to consist of a summit zone (if it exists) corresponding to the brittle fracture, a transition zone in which both phenomena can be mixed, and the plastic yield boundary.

Experience indicates that when the spherical part of the stress-tensor is relatively large the material is no longer in the brittle-fracture range, and ductile rupture occurs. Moreover, experiments show that for materials composed of a single phase (this excludes porous materials—see [39, 40]), when the stress state is isotropic, however high the pressure, there is no plastic yielding. The loading surface is thus open (with the usual meaning of the word) in the direction of the isotropic pressures.[1]

---

[1] For porous materials, the elastic range is closed in all directions. The utilization for soils, which are granular materials, of Coulomb's yield criterion or of other criteria of the 'open' intrinsic curve type corresponds to a quite justified approximation for the range of actual stresses.

For ductile materials, there is no brittle fracture, no summit zone, and the loading surface is open towards the two directions corresponding to the isotropic pressures and tensions.

2.9 Anisotropic results

The previous paragraphs have presented the main yield criteria used for isotropic materials. Real materials are often anisotropic (metals, soils or rocks—see, for example, [1] and [53] for both latter types of materials).

Various authors have dealt with the problem of anisotropic plasticity including Hill [19–21], who obtained interesting results for a metallurgy problem, Caquot and Kerisel [5], who introduced the notion of a tensor or anisotropy and the authors of [2–4, 22], etc.[1]

3. Plastic Deformation

As stated in Section 1, passing the yield point or crossing the yield surface corresponds to the appearance of permanent deformations. Plastic deformation is defined as follows. In the normal case, where the elasticity of the material concerned is linear and not modified by the plastic yielding, the plastic deformation is the permanent deformation after complete unloading of the material, assuming the unloading to be elastic. After a given degree of work-hardening it may happen that the real total unloading cannot be completely elastic, so that the zero loading no longer belongs to the elastic range of the element. In this case, the plastic deformation is the permanent deformation obtained after a virtual entirely elastic total unloading. (See Figure I.4 for the case of a uniaxial loading process.)

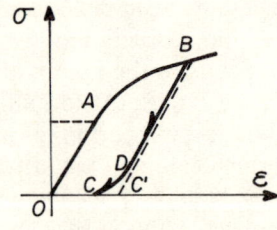

Figure I.4

The exact definition of the plastic deformation, valid in all cases, refers to strain-rates. If v is defined as the strain-rate tensor then, at each moment of the loading path, the total strain rate v_{ij} (the result of the test) and the elastic strain

---

[1] The loading functions proposed by these authors are expressed using Eulerian stress-tensor; as already stated, this is approximate.

rate v_{ij}^e are known (an infinitesimal unloading is sufficient). The plastic strain rate is the difference between these two; i.e.

$$v_{ij} = v_{ij}^e + v_{ij}^p \qquad (12)$$

The plastic deformation is the integral of v_{ij}^p along the loading path.[1]

4. The Origin of the Plastic Deformation

The plastic deformation of solids is explained by Mandel [33] as follows. In the solid, which is an assembly of crystalline grains, plastic deformation can result from either of two phenomena.

(1). The relative motion of the grains themselves. In the case of soils, the irreversibility of the deformation is explained by the friction between the grains.

(2). The permanent deformations of the grains. This applies to metals. The deformations arise by sliding in the crystal lattice of the grain, along the atomic planes. The phenomenon is explained by the theory of dislocations (see for example [12]).

5. Flow Rule

The previous definitions make it possible to determine the onset of the plastic deformation, and its nature. In order to have a complete knowledge of the constitutive law of plasticity, the questions of mechanism and magnitude must also be considered. The answer constitutes the flow rule.

Figures I.1 and I.4 clearly show that the constitutive law cannot lead (as is the case in elasticity) to a unique relation between the actual state of stresses and the actual strain. (Thus, the points O and C correspond to the same loading state but to different deformations). The actual deformation depends on the loading path followed before reaching the actual loading state.[2]

If the actual state of stress and the work-hardening state, i.e. the actual values of the work-hardening parameters representing the loading history, are known, then the increment of deformation may be determined from the increment of

---

[1] Rigorously speaking, Green's strain-tensor ought to have been used; moreover, the notion of unloaded state or neutral state is fundamental here; the choice of this reference configuration, investigated in [29, 43, 52] for the isotropic case, was carefully looked at by Mandel [36, 37] in more general cases.

[2] It does not depend on time since viscosity phenomena have been excluded. The time taken into account in static or quasistatic plasticity is therefore no more than a kinematic parameter, the scale of which can be modified as desired. Therefore time can play no role whatever in $A_{ij,hk}$ in equations (14) and (15). In actual fact, a material's irreversible deformations always contain a viscous part, and physical time must be present in the flow rule, and even in the very notion of yield point, which depends on the loading speed. (This is indicated in particular by the theory of dislocations for metals [58].) The study of visco-plasticity, after a purely phenomenological approach that led to limited results, now seems to be leading to a more fundamental theory. (Cf. [15] where there are also references to former studies of the subject).

Obviously enough, the utilization of the model of plastic behaviour without viscosity will be justified for phenomena in which viscosity does not preponderate.

stress. (In the uniaxial case of Figures I.1a and I.4, ε plays the role of the work-hardening parameter[1]). Such a stress–strain relation is termed an incremental constitutive law.

In the general case of multiaxial loading the tensorial formula is given by

$$d\varepsilon = \mathscr{B}(\sigma, E, d\sigma)$$

which is independent of $\dot{\sigma}$ because of the absence of viscosity. It assumes the differential form

$$d\varepsilon = \quad (\sigma, E) \, d\sigma \qquad (13)$$
$$d\varepsilon_{ij} = \mathscr{B}_{ij,hk}(\sigma, E) \, d\sigma_{hk}$$

In terms of the strain and stress rates, the relation becomes

$$\mathbf{v} = \mathscr{B}(\sigma, E, \dot{\sigma})$$

which must be homogeneous with respect to (kinematic) time, so that

$$\mathbf{v} = \mathscr{B}(\sigma, E)\dot{\sigma}$$

In the decomposition of Section 3, $(d\varepsilon^e)_{ij}$ follows, by definition, the elastic constitutive law

$$d\varepsilon^e = \Lambda \, d\sigma$$

If A is defined by

$$\mathscr{B} = A + \Lambda$$

then

$$(d\varepsilon^p)_{ij} = A_{ij,hk}(\sigma, E) \, d\sigma_{hk} \qquad (14)$$

or, in terms of strain- and stress-rates,

$$v_{ij} = A_{ij,hk}(\sigma, E)\dot{\sigma}_{hk}.^2 \qquad (15)$$

Because of the material behaviour being different during 'loading' and 'unloading' (irreversiblity characteristic of plasticity), $A_{ij,hk}$ will assume two different expressions.

These expressions may be specified as follows. According to the result of the tests indicated in Section 1, the plastic deformation of a work-hardening material for which the yield criterion is satisfied occurs only if loading continues (Figure I.5).[3] If 'the differential of f at constant work-hardening' is denoted by

$$d_E f = \frac{\partial f}{\partial \sigma_{hk}} \, d\sigma_{hk} \qquad (16)$$

---

[1] For a perfectly plastic material (Figure I.1b), when $\sigma = \sigma_A$ and $d\sigma = 0$, $d\varepsilon$ is undetermined. We can say that any material exhibits some work-hardening, little though it be, that makes this result physically insignificant. This work-hardening is, in some cases, small enough to be neglected in problems likely to be simplified by so doing.

[2] Strictly speaking, objective definitions (derived from Truesdell, for instance see [33]), must be taken for $d\sigma$ and σ in formulae (14), (15).

[3] It is known that in this case the yield surface, according to the very definition of work-hardening, is extended with the load point; whence $f(\sigma + d\sigma, E + dE) = 0$.

this deformation condition, simple in the case of loading with only one parameter, may be expressed by $d_E f > 0$. Then

$$(d\varepsilon^p)_{ij} \neq 0$$

is given by equation (14) where $A \neq 0$.

If the load point remains on the current loading surface, i.e. if

$$d_E f = 0$$

then the plastic deformation is zero:

$$(d\varepsilon^p)_{ij} = 0$$

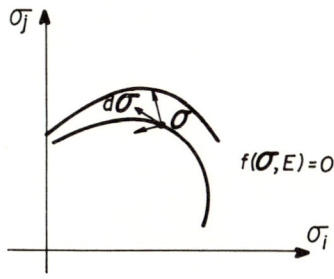

Figure I.5

Finally, during *unloading*, i.e. if

$$d_E f < 0$$

there is no plastic deformation

$$(d\varepsilon^p)_{ij} = 0.[1]$$

Moreover, the passage is continuous from one case to the other; i.e.

if $d_E f \searrow 0$, then $(d\varepsilon^p)_{ij} \to 0$

Therefore, each linear form $(d\varepsilon^p)_{ij}$ in $d\sigma_{hk}$ of equation (14) is zero when the linear form $d_E f$ in $d\sigma_{j,k}$ of equation (16) is zero. These forms are therefore dependent and

$$d\varepsilon_{ij}^p = H_{ij}(\sigma, E) \, d_E f \qquad (17)$$

Thus, the increment $d\sigma$ of *tensor* σ is present in $d\varepsilon^p$ only via the *scalar* $d_E f$.

Because of this, the principal directions of the plastic deformation increment as well as the ratios of its principal values do not depend on the increments of stresses. This fact was verified by Morrisson and Shepherd's experiments (1950) on the tension and torsion of metal wires.

In the case of an isotropic material subject to isotropic work-hardening **H** is an

---

[1] Also, there is no modification at the yield surface.

isotropic tensorial function of $\boldsymbol{\sigma}$ and has therefore the same principal directions as $\boldsymbol{\sigma}$. It follows that $d\boldsymbol{\varepsilon}^p$ and $\boldsymbol{\sigma}$ have identical principal directions.[1]

To summarize, for a work-hardening material, the general argument makes it possible to specify the stress–strain relation for the plastic deformation in the form

$$(d\varepsilon^p)_{ij} = H_{ij}(\boldsymbol{\sigma}, E)\frac{\partial f}{\partial \sigma_{hk}}d\sigma_{hk} \quad \text{if} \begin{cases} f(\sigma_{hk}, E) = 0 \\ \dfrac{\partial f}{\partial \sigma_{hk}}d\sigma_{hk} \geq 0 \end{cases}$$

$(d\varepsilon^p)_{ij} = b$ in the other cases

No more can be said about the flow rule without any complementary hypothesis.

For perfectly plastic materials, the stress–strain relation is simply

$$(d\varepsilon^p)_{ij} \neq 0 \quad \text{or} \quad =0 \quad \text{if} \quad f(\sigma_{ij}) = 0 \quad \text{and} \quad \frac{\partial f}{\partial \sigma_{ij}}d\sigma_{ij} = 0$$

$(d\varepsilon^p)_{ij} = b$ in the other cases

Moreover, if the material is isotropic, the tensors $\mathbf{v}^p$ and $\boldsymbol{\sigma}$ have identical principal directions.

6. Principle of Maximum Plastic Work

It is often assumed that materials obey Hill's principle of maximum plastic work (1950), which can be stated as follows.

If $\boldsymbol{\sigma}$ is a plastically admissible stress tensor, i.e. such that $f(\boldsymbol{\sigma}) < 0$, $\mathbf{v}^p$ a strain-rate tensor corresponding to this stress-state according to the plastic constitutive law, and $\boldsymbol{\sigma}^*$ is another plastically admissible stress tensor ($f(\boldsymbol{\sigma}^*) \leq 0$), then

$$(\sigma_{ij} - \sigma_{ij}^*)v_{ij}^p \geq 0 \tag{18}$$

which may also be written

$$(\boldsymbol{\sigma} - \boldsymbol{\sigma}^*)\mathbf{v}^p \geq 0$$

Geometrically, the inequality (18) shows that if tensors $\mathbf{v}^p$ and $\boldsymbol{\sigma}$ (both symmetrical), are represented by vectors of a six-dimensional space (see [32]), the scalar product $\sigma^*\sigma \cdot v^p$ is ≥ 0 if $\boldsymbol{\sigma}$ is on the surface $f(\boldsymbol{\sigma}) = 0$ and if $\boldsymbol{\sigma}^*$ is not outside it. This implies the *convexity* of the loading surface and $\mathbf{v}^p$ is an outward normal to this surface at point $\boldsymbol{\sigma}$.[2]

Mathematically, taking into account the symmetry of tensors $\mathbf{v}^p$ and $\boldsymbol{\sigma}$, and the fact that f is symmetrical in σ_{ij} and σ_{ji}, the following argument applies. If the surface is regular at point $\boldsymbol{\sigma}$, then

$$\mathbf{v}^p = \lambda\frac{\partial f}{\partial \boldsymbol{\sigma}}, \qquad \lambda \geq 0 \tag{19}$$

---

[1] More precisely, $\mathbf{H}$ and $\boldsymbol{\sigma}$, $d\varepsilon^p$ and $\boldsymbol{\sigma}$, $\mathbf{v}^p$ and $\boldsymbol{\sigma}$ have at least one system of principal directions in common.

[2] Figure I.6 was drawn for the case in which there is isotropy, aiming at greater simplicity.

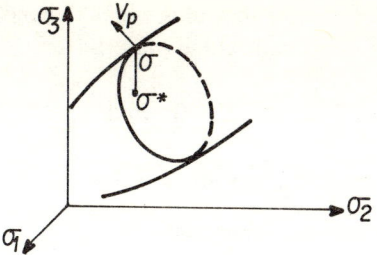

Figure I.6

or, in order to include the case of a conical point,

$$\mathbf{v}^p \in \lambda \partial f(\boldsymbol{\sigma}), \qquad \lambda \geq 0.^1 \tag{20}$$

f is therefore the *plastic potential*.

With a work-hardening material, the principle of maximum plastic work may be assumed at each stage of the work-hardening.

In the case of metals, the sliding along atomic planes is governed by Schmid's law: the sliding occurs if the shear stress on the plane in the direction of the sliding attains a critical value. The principle of maximum work can be proved as a consequence of Schmid's law (see [33], p. 720).

It must be noted that some authors (e.g. [27]) prefer to accept Drucker's 'quasi-thermodynamic' postulate [8] as a basis. As the mathematical equivalence of Drucker's and Hill's formulation can be proved, the above results concerning the yield criterion and the flow rule still hold.

Yet other authors prefer to assume *a priori* both the properties of *convexity* of the criterion and of *normality* of the flow rule, rather than to use Hill's principle or Drucker's postulate, whose physical foundations are far from obvious.

7. Consequences of the Constitutive Law

Assuming the validity of the principle of maximum plastic work (i.e. the convex loading function to be also the plastic potential) leads, in the case of work-hardening, to

$$\left. \begin{array}{l} v_{ij}^p = \dfrac{1}{M} \dfrac{\partial f}{\partial \sigma_{ij}} \left(\dfrac{\partial f}{\partial \sigma_{hk}} \dot{\sigma}_{hk} \right) \quad \text{if} \quad f = 0 \quad \text{and} \quad \dfrac{\partial f}{\partial \sigma_{hk}} \dot{\sigma}_{hk} \geq 0 \\ M \geq 0 \\ v_{ij}^p = 0 \text{ in the other cases} \end{array} \right\} \tag{21}$$

---

[1] $\partial f(\boldsymbol{\sigma})$ is the subdifferential of function f at point $\boldsymbol{\sigma}$. $\lambda \partial f(\boldsymbol{\sigma})$, $\lambda \geq 0$, is the cone of outward normals to f at this point.

where M is the work-hardening modulus which depends *a priori* on the loading history. In the case of a non-hardening material

$$\left.\begin{array}{l} v_{ij}^p = \lambda \dfrac{\partial f}{\partial \sigma_{ij}} \quad \text{if} \quad f = 0 \quad \text{and} \quad \dfrac{\partial f}{\partial \sigma_{hk}} \dot{\sigma}_{hk} = 0 \\[2mm] \lambda \geqslant 0, \text{ arbitrary positive factor} \\[2mm] v_{ij}^p = 0 \text{ in the other cases} \end{array}\right\} \quad (22)$$

For such material, the flow rule is said to be *associated*, and the material to be *standard* (Radenkovic [45, 46]).

As an example, for a Von Mises standard material without work-hardening, the flow rule is

$$\mathbf{v}^p = \lambda \boldsymbol{\varepsilon}, \quad \lambda \geqslant 0 \quad \text{if} \quad J_2 = k^2 \quad \text{and} \quad \dot{J}_2 = 0$$
$$\mathbf{v}^p = 0 \text{ in the other cases.}$$

For a Tresca material, the flow rule, referred to the principal axes, is

$$v_1^p = \lambda, v_2^p = 0, v_3^p = -\lambda, \lambda \geqslant 0 \qquad \text{if} \quad \sigma_1 > \sigma_2 > \sigma_3$$
$$\text{and} \quad \sigma_1 - \sigma_3 = 2k$$
$$v_1^p = \lambda + \mu, v_2^p = -\mu, v_3^p = -\lambda, \lambda \geqslant 0, \mu \geqslant 0 \quad \text{if} \quad \sigma_1 - \sigma_3 = \sigma_1 - \sigma_2 = 2k$$

etc.

8. Validity of the Principle of Maximum Plastic Work for the Case of Soils

From experimental results, the principle of maximum plastic work appears to be valid for materials whose yield criterion is independent of the average pressure $(-\sigma_{ii}/3)$. This is the case for ductile metals, and also for clays with $\phi = 0$ (undrained soil). For these plastic deformation occurs without volume change.

Nevertheless, investigations on such materials have been carried out using Tresca's criterion and the flow rule associated with a Mises' potential (e.g. [21]).

For a soil following Coulomb's criterion (10), the principle of maximum plastic work would lead to the following flow rule:

$$\left.\begin{array}{l} v_1^p = \lambda(1 + \sin \phi) \\ v_2^p = 0 \\ v_3^p = -\lambda(1 - \sin \phi) \\ \lambda \geqslant 0 \end{array}\right\} \quad (23)$$

A change of volume would result from the plastic deformation, equal to

$$\dot{\theta} = v_1^p + v_3^p = 2\lambda \sin \phi = \left[1 - \tan^2\left(\frac{\pi}{4} - \frac{\phi}{2}\right)\right] v_1^p = \sin \phi (v_1^p - v_3^p)$$

Such a significant volume change is not in agreement with experiments. Thus, the principle of maximum plastic work is not admissible for soils obeying Coulomb's criterion ($\phi \neq 0$).

Some authors (see, for example, [17]) have adopted for soils the hypothesis that the plastic deformation occurs without volume change; i.e. $v_{ii}^p = 0$. As the material is isotropic, tensors $\mathbf{v}^p$ and $\boldsymbol{\sigma}$ must have identical principal directions. The flow rule of Tresca's standard material (i.e. obeying the principle of maximum plastic work) may be relevant in this case, i.e.

$$\left. \begin{array}{l} v_1^p = \lambda \\ v_2^p = 0 \\ v_3^p = -\lambda \\ \lambda \geqslant 0 \end{array} \right\} \qquad (24)$$

According to [23], this hypothesis is admissible for Coulomb-type soils if they are at the critical void ratio for the established plastic deformation [24].

$$\left. \begin{array}{l} v_1^p = \lambda(1 + \sin v) \\ v_2^p = 0 \\ v_3^p = -\lambda(1 - \sin v) \\ \lambda \geqslant 0 \end{array} \right\} \qquad (25)$$

in which the angle v is the so-called angle of dilation of the element. (v is positive if there is dilation.)

In fact, angle v must vary as the plastic deformation of the element proceeds, and therefore it would be difficult to use rule (25). However, according to [23], satisfactory results are obtained by considering an idealization of the material with v constant, $0 \leqslant v \leqslant \phi$, as proposed in [18], [45], and used, for example, in [50].

Thus, it may be considered that soils have a Coulomb yield function (angle ϕ) and a Coulomb plastic potential (angle $v \neq \phi$) different from the yield function.

9. Final Remarks

It must be emphasized that the definition of plastic behaviour for a material consists of two parts:

(1). The yield criterion,
(2). The flow rule.

The yield criterion (or, more accurately, the yield function) intervenes in the flow rule only for a work-hardening material. In this case it is used to 'measure' the intensity of load supported by the element of material. Only when the hypothesis of the principle of maximum work is effected does the yield criterion also define the flow rule.

These aspects have not always been well distinguished, as indicated by Roscoe [49], for 'Mohr–Coulomb's rupture criterion'. This point will be dealt with when studying plane plastic flow.

10. Application of the Plasticity Model to Soil Behaviour

The prediction of soil behaviour via plasticity models does raise numerous problems. However, the results obtained by this means have proved of interest for many engineering problems. This point will not be discussed here, but reference will be made to relevant literature.

As stated in Section 9, the conditions for application of the yield-point and yield-criterion concepts, concerning their significance for soils on the one hand and the question of the flow rule on the othe, must be examined. For this kind of analysis reference will be made, for example, to [47] and [51].

With respect to the yield criterion the experiments performed with true triaxial apparatus [14, 28] indicate that the Coulomb criterion in the form of an intrinsic curve models true behaviour fairly closely.

Regarding the flow rule, there has been much discussion in the case of soils. Various authors [7, 13, 25, 26, 31, 34, 38, 54] have attempted to determine a flow rule by consideration of the kinematics of the granular medium, the basic concept being that the plastic deformation takes place by sliding. Others [15, 17, 18, 41] preferred the approach based on the notion of a plastic potential different from the yield criterion, and have proposed flow rules such as those indicated in Section 8.

Finally, we must note that Mandel's works [36, 37], which introduced a *director trihedron* for each element of the plastic medium, might be useful for granular media. In particular, light could be shed on the problems connected with the *non-coaxiality* of tensors v^p and σ for isotropic materials.

References

[1] J. Biarez (1962) Contribution à l'étude des propriétés mécaniques des sols et des matériaux pulvérulents, Thesis Dr. Sc., Grenoble (1962).

[2] J. P. Boehler and A. Sawczuk (1970) Equilibre limite des sols anisotropes, *J. Mec.*, **9**, pp. 5–33.

[3] J. R. Booker and E. H. Davis, (1972) A general treatment of plastic anisotropy under conditions of plane strain, *J. Mech. Phys. Solids*, **20**, n° 4, pp. 239–250.

[4] J. Boschat and D. Radenkovic, (1962) Une généralisation de la loi limite de Tresca aux matériaux anisotropes, *Z. Ang. Math. Mech.*, **42**, pp. 90–91.

[5] A. Caquot and J. Kerisel (1956) *Traité de Mécanique des sols*, 3rd. ed., Gauthier-Villars, Paris.

[6] A. Drescher (1972) Some remarks on plane flow of granular media, *Archives of Mechanics*, **24**, 5–6, pp. 837–848.

[7] A. Drescher and G. de Josselin de Jong (1972) Photoelastic verification of a mechanical model for the flow of a granular material, *J. Mech. Phys. Solids*, **20**, pp. 337–351.

[8] D. C. Drucker (1951) A more fundamental approach to plastic stress–strain relations, *Proc. 1st U.S. Nat. Congr. Appl. Mech.*, p. 487.

[9] D. C. Drucker (1956) On uniqueness in the theory of plasticity, *Quart. Appl. Math.*, **26**, n° 1, pp. 35–42.
[10] D. C. Drucker (1961) On stress–strain relations for soils and load carrying capacity, *Proc. 1st Int. Conf. on the Mechanics of Soil Vehicle Systems*, Torino, Italy, 12–16 June 1961.
[11] D. C. Drucker and W. Prager (1952) Soil Mechanics and Plastic Analysis or or Limit Design. *Quart. Appl. Math.*, **10**, pp. 157–165.
[12] J. Friedel (1966) *Dislocations*, Pergamon Press, Oxford.
[13] G. A. Geniev (1958) *Problems of Dynamics of Granular Media*, (in Russian), Gostekhizd, Moscow.
[14] M. Goldscheider and G. Gudehus (1973) Some sectionally proportional rectilinear extension tests on dry sand, *Proc. Symp. on the Role of Plasticity in Soil Mechanics*, Cambridge (G.B.), 13–15 Sept. 1973, pp. 56–66.
[15] G. Gudehus (1973) Elastoplastische Stoffgleichungen für trockenen Sand, *Ingenieur Archiv*, **42**, pp. 151–169.
[16] P. Habib (1953) Etude de l'orientation du plan de rupture et de l'angle de frottement interne de certaines argiles, *Proc. 3rd Int. Conf. Soil Mech.*, Zürich, **1**, pp. 28–31.
[17] J. Brinch Hansen (1953) *Earth Pressure Calculation*, Danish Technical Press, Copenhagen.
[18] Bent Hansen (1958) Line ruptures regarded as narrow rupture zones, *Proc. Conf. sur les Problèmes de poussées des terres*, Bruxelles, **1**, pp. 39–49.
[19] R. Hill (1948) A theory of the yielding and plastic flow of anisotropic metals. *Proc. Roy. Soc. A*, **193**, pp. 281–289.
[20] R. Hill (1949) The theory of plane plastic strain for anisotropic metals. *Proc. Roy. Soc. A*, **198**, pp. 428–437.
[21] R. Hill (1950) *The Mathematical Theory of Plasticity*, Clarendon Press, Oxford.
[22] H. Igaki, M. Sugimoto and K. Saito (1970) Anisotropic yield criterion under the maximum shear stress theory. *J.S.M.E.*, **13**, n° 61, pp. 825–836.
[23] R. G. James and P. L. Bransby (1971) A velocity field for some passive earth pressure problems. *Géotechnique*, **21**, n° 1, pp. 61–83.
[24] A. W. Janike and R. T. Shield (1959) On the plastic flow of Coulomb solids beyond original failure. *J. Appl. Mech., Trans. ASME*, **26**, n° 4, pp. 599–602.
[25] G. de Josselin de Jong (1964) Lower bound collapse theorem and lack of normality of strain-rate to yield surface for soils, *Proc. I.U.T.A.M. Symp. on Rheology & Soil Mechanics*, Grenoble, pp. 69–75.
[26] G. de Josselin de Jong (1971) The double sliding, free-rotating model for granular assemblies, *Géotechnique*, **21**, n° 2, pp. 155–163.
[27] W. T. Koiter (1960) General theorems for elastic-plastic solids. *Progress in Solid Mechanics*, Ed. Sneddon and Hill, Vol. 1, pp. 165–221., North-Holland Pub. Co., Amsterdam.
[28] P. V. Lade (1973) Discussion, session II. *Proc. Symp. on the role of Plasticity in Soil Mechanics*, Cambridge (G.B.), 13–15 Sept. 1973, pp. 129–135.
[29] E. H. Lee (1969) Elastic-plastic deformation at finite strains, *J. Appl. Mech., Serie E*, **36**, n° 1, pp. 1–6.
[30] J. Legrand (1969) *Cours de Mécanique des Sols*, E.N.P.C., Paris.
[31] J. Mandel (1947) Sur les lignes de glissement et le calcul des déplacements dans la déformation plastique. *C.R.Ac.Sc.*, Paris, t.225, pp. 1272–1273.
[32] J. Mandel (1964) Contribution théorique à l'étude de l'écrouissage et des lois de l'écoulement plastique, *Proc. 11th Int. Congr. Appl. Mech.*, Münich, pp. 502–509.
[33] J. Mandel (1966a) *Mécanique des Milieux Continus*, Vol. II, Gauthier-Villars, Paris.
[34] J. Mandel (1966b) Sur les équations d'écoulement des sols idéaux en déformation plane et le concept du double glissement, *J. Mech. Phys. Solids*, **14**, pp. 303–308.
[35] J. Mandel (1969) *Cours de Science des Matériaux*, Ecole Nationale Supérieure des Mines de Paris.

[36] J. Mandel (1971a) Sur la décomposition d'une transformation élastoplastique, *C.R.Ac.Sc., Paris*, **272**, A, pp. 276–279.

[37] J. Mandel (1971b) *Plasticité Classique et Viscoplasticité*. C.I.S.M., Udine (Italy), Springer-Verlag.

[38] G. Mandl and R. Fernandez-Luque (1970) Fully developed plastic shear flow of granular materials. *Géotechnique*, **20**, n° 3, pp. 277–307.

[39] P. Morlier (1970a) Plasticité et écrouissage d'un métal fritté. *Mém. Sc. Rev. de Métallurgie*, **67**, No. 6, pp. 401–412.

[40] P. Morlier (1970b) Comportement des roches sous contrainte en fonction de leur teneur en eau, *La Houille Blanche*, **5**, pp. 471–475.

[41] Z. Mroz (1963) Non associated flow-laws in Plasticity, *J. Mécanique*, **2**, No. 1, pp. 21–42.

[42] V. N. Nikolaevskii (1971) Governing equations of plastic deformation of a granular medium, *J. Appl. Math. & Mech.* (trans. P.M.M.), **35**, No. 6, pp. 1017–1029.

[43] M. Piau (1970) Description du comportement d'une classe de milieux continus..., *J. Mec.*, **9**, No. 3, pp. 375–401.

[44] W. Prager (1951) *Theory of Perfectly Plastic Solids*, John Wiley, New York.

[45] D. Radenkovic (1961) Théorèmes limites pour un matériau de Coulomb à dilatation non standardisée, *C.R.Ac.Sc., Paris*, **252**, pp. 4103–4104.

[46] D. Radenkovic (1962) Théorie des charges limites, *Séminaire de Plasticité*, Ed. J. Mandel, pp. 129–142.

[47] D. Radenkovic (1972) Equilibre limite des milieux granulaires. Modeles de comportement rigide-plastique, *Plasticité et Viscoplasticité*, Ed. D. Radenkovic and J. Salençon, Ediscience, Paris, 1974, pp. 379–394.

[48] K. H. Roscoe et al. (1967) Principal axes observed during simple shear of a sand, *Proc. Geot. Conf., Oslo*, pp. 231–238.

[49] K. H. Roscoe (1970) 10th Rankine Lecture: the influence of strains in soil mechanics, *Géotechnique*, **20**, No. 2, pp. 129–170.

[50] J. Salençon (1966) Expansion d'une cavité dans un milieu élastoplastique, *Ann. Pts. Ch.*, 1966, **3**, pp. 175–187.

[51] J. Salençon (1974) Plasticité pour la Mécanique des Sols. C.I.S.M., Rankine Session, July 1974, Udine (Italy).

[52] F. Sidoroff (1970) Quelques réflexions sur le principe d'indifférence matérielle, pour un milieu ayant un état relàché, *C.R.Ac.Sc., Paris*, **271**, A, pp. 1026–1029.

[53] P. M. Sirieys (1966) Contribution à l'étude des lois de comportement des structures rocheuses. Thesis Dr. Sc., Grenoble.

[54] A. J. M. Spencer (1964) A theory of the kinematics of ideal soils under plane strain conditions, *J. Mech. Phys. Solids*, **12**, No. 5, pp. 337–351.

[55] P. Stutz (1972) Comportement élastoplastique des milieux pulvérulents, *Plasticité et Viscoplasticité*, Ed. D. Radenkovic and J. Salençon, Ediscience, Paris, 1974, pp. 395–420.

[56] A. S. Wineman and A. C. Pipkin (1964) Material symmetry restrictions on constitutive equations, *Arch. Rat. Mech. Anal.*, **16**, pp. 184–214.

[57] R. N. Young and E. McKyes (1971) Yield and failure of a clay under triaxial stresses, *J. Soil Mech. & Found. Div., Proc. ASCE*, **97**, SM1, pp. 159–176.

[58] J. Zarka (1970) Sur la viscoplasticité des métaux, *Mém. Art. Fr. (Sc. Tech. Arm.)*, fasc. 2, 1970, pp. 223–292.

[59] H. Zielger (1969) Zum plastischen potential in der Bodenmechanik, *Z.A.M.P.*, **20**, pp. 659–675.

CHAPTER II

A short survey of the problems of elasto-plasticity

1. Introduction

The so-called problems of elasto-plasticity are problems in which the constitutive equation adopted for material behaviour is of the type indicated in Chapter I. The deformation consists of an elastic and a plastic part. For example, in the case of an isotropic linear elastic material obeying Mises' yield criterion with work-hardening and the principle of maximum plastic work, the constitutive equation is:

$$\left.\begin{array}{l} v_{ij} = \dfrac{1+v}{E}\dot{\sigma}_{ij} - \dfrac{v}{E}\dot{I}_1\delta_{ij} + \psi s_{ij} \\[6pt] I_1 = \sigma_{ii} \\[4pt] \text{where}\quad \psi = g(J_2, J_3, E)\dot{J}_2 \geqslant 0 \end{array}\right\} \quad \text{if } \dot{J}_2 \geqslant 0 \quad \text{and} \quad J_2 = k^2(E)$$

$$\left.\begin{array}{l} v_{ij} = \dfrac{1+v}{E}\dot{\sigma}_{ij} - \dfrac{v}{E}\delta_{ij}\dot{I}_1 \end{array}\right\} \quad \text{in the other cases} \tag{1}$$

which is termed the *Prandtl–Reuss law*.

Problems of this nature arise as soon as plasticity (without viscosity) becomes significant in the deformation of materials. As will be seen, they are difficult to solve, and few can be solved explicitly. It is not intended to examine them in detail, but only to investigate the form in which they appear and to make some general observations. (Reference may be made to classical text books, particularly [6], for a more detailed presentation: theorems of uniqueness of the solution, minimum principles, etc.)

2. Statement of the Problem. Method of Solution

The objective is to determine the actual state of stress and strain of an elasto-plastic system. The problem is to be solved incrementally following the loading path. This differential approach is dictated by the incremental form of the constitutive equation. In general, the state of stress and strain within the system depends on the *loading path* followed to attain the actual loading.

The incremental method of solution is as follows. At any time t the stress and displacement fields are known, and by using the equilibrium equations, the constitutive equation. ((1) for instance), and the boundary conditions imposed on the velocities and on the rates of the imposed forces, it is possible to determine the stress-rate field $\dot{\sigma}_{ij}$ and the velocity field u_i throughout the whole solid. It is then possible to determine the stress and displacement fields at time $(t + dt)$.

It must be emphasized that for each time increment, the boundaries between the elastic and plastic zones vary and must be redetermined. Points situated within the elastic zones ($f(\sigma) < 0$) during the previous time step now lie within plastic zones, and points situated within the plastic zones during the previous step ($f(\sigma) = 0$) now lie within elastic zones (an occurrence known as 'local unloading'). This movement of state boundaries is the main source of practical difficulties.

Only a few problems, in which important simplifications arise due to geometrical symmetry, can be solved explicitly by analytical means (see [12] and [13]). Usually, numerical methods (based on minimum principles) are used. These involve discretization in space (for example, by means of the finite element method [4], [15], [17] etc.), as well as in time. The approximation in time may lead to problems of convergence of the numerical scheme, as indicated in [9]. Numerical methods have been used (assuming small deformations) to solve problems concerning Tresca or Mises materials with positive work-hardening (or even negative), or Coulomb material using either the associated flow rule, or one of the rules indicated in Chapter I.

3. Behaviour of a System Subject to a Loading Process

3.1. Loading depending on one parameter

Let us consider an elasto-plastic system, constituted by a solid or a set of solids. This system is subjected to loads proportional to a parameter Q and undergoes the following loading process: starting from the natural state, with zero stresses for $Q = 0$, Q is made to increase.

During the solution of this elasto-plastic problem, assuming small deformations (i.e. *changes of geometry* being neglected), the following stages are met.

Firstly, as long as Q is less than a value Q_0, there are no plastic zones and we have $f(\sigma) < 0$ everywhere.

For $Q = Q_0$ plastic zones appear and the yield state is reached at a point (or several points simultaneously). Q_0 is termed the initial *elastic limit* of the system.

As Q increases beyond Q_0, the plastic zones spread. Plastic deformations are possible within these zones, but they are contained by the deformations of the elastic zones. This means that, at any point within the plastic zones, the plastic deformation is limited for each value of Q, whether the material is work-hardening or perfectly plastic, owing to the continuity of the material through the boundary between the elastic and plastic zones.

Finally, when Q reaches the value Q_1, the plastic zones have developed sufficiently to make *uncontained plastic flow* possible, i.e. plastic deformations are no longer contained by elastic zones.

From this point on, the behaviour of the system depends upon whether the material is work-hardening or perfectly plastic. In the latter case, with geometry changes being neglected, the deformation in the plastic zones becomes unlimited for $Q = Q_1$. This unlimited deformation of a part of the system, under constant loading, is known as collapse. The value Q_1 is termed the *limit load* of the system. In the case of a work-hardening material the deformation in the plastic zones is still determinate. With geometry changes being neglected, the loading must increase before deformation can continue.

As an example, the plane problem of a uniform pressure with variable intensity Q, applied to the surface of a half-plane consisting of a perfectly plastic Tresca material,[1] may be considered. Q_0 is found to be equal to πk, and the plastic zone then consists of the semi-circle with diameter $A'A$ (Figure II.1a); for the limit load Q_1, found to be equal to $(\pi + 2)k$, the plastic zone may be represented schematically as in Figure II.1b and the material can flow towards the surface in both directions.

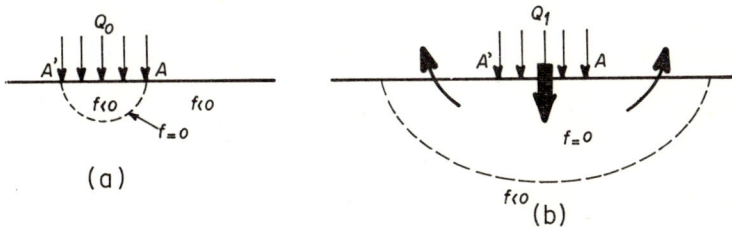

Figure II.1

3.2 Loading depending upon several parameters

For the case of a load which is dependent on a finite number of parameters Q_i,[2] the concepts of Section 3.1 are applied as follows.

Let us denote by **Q** the point with coordinates $Q_1, Q_2, \ldots Q_n$ in a n-dimensional space (as no confusion may arise **Q** will also denote the vector with the same coordinates in a b-dimensional linear space).

The *initial elastic domain* of the system is then defined as the domain for the loading-point **Q**, in which the system remains elastic along any loading-path starting from the natural state with zero-stresses for **Q** = 0. This domain is limited by the *initial elastic boundary*. As a consequence of the reversibility of the

---

[1] This material is homogeneous, with linear isotropic elasticity.
[2] For example subject to n forces or active pressures the respective intensities of which are $Q_i (i = 1, \ldots, n)$. The notion of loading parameters will be explained in detail in Chapter III, so as to include more difficult cases of boundary conditions.

elastic deformations, that boundary is proved to be the locus of the points Q_0 defined on each of the above-mentioned loading-paths by the first appearance of plasticity in the system. It is a convex surface for the hypothesis of a convex yield function and of linear elasticity.

If the load is increased beyond this boundary, the elastic domain of the system changes and an *actual elastic boundary* of the system, limiting the actual elastic domain is defined. The domain is, so to speak, hauled by the loading point Q. This is the 'work-hardening of the system' due to the work-hardening of the plastic elements of the system and the incompatibility of the plastic deformations (see [5]). Obviously the actual elastic boundary depends on the loading path followed.

The load being increased along a given loading path a point Q_1 is reached for which uncontained plastic flow becomes possible. The locus of points Q_1 corresponding to all the loading paths is termed the *yield boundary of the system*.

As an example, Figure II. 2b taken from reference [5], shows the different boundaries for the structure represented in Figure II.2a.

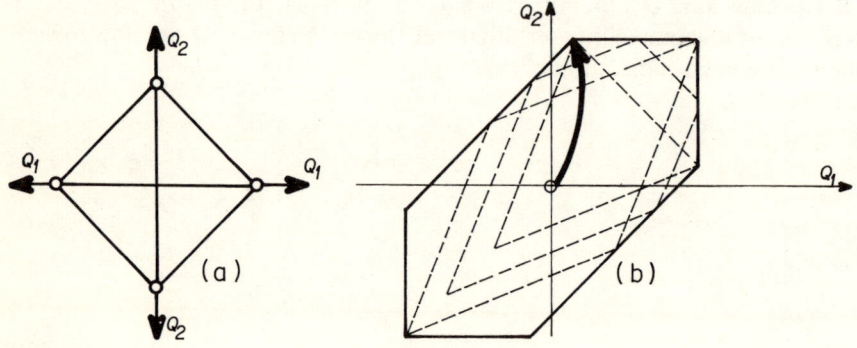

Figure II.2

4. Importance of the Geometry Changes

The description of the behaviour of a system subject to a loading process, given in Section 3.1, is realistic only if the geometry changes are actually negligible.

4.1 Geometry changes during uncontained plastic flow

Firstly, it is obvious that if the stage of uncontained plastic flow has been reached then the deformations are no longer limited to the order of elastic deformations and it may now be necessary to take geometry changes into account. This has the following consequences: for a perfectly plastic material, either the load necessary to continue the deformation increases endlessly, or it increases and reaches a maximum, thus giving rise to instability and collapse.

However, in both cases Q_1 does have a practical meaning as it corresponds approximately to the appearance of inadmissible deformations in the structure.[1] For a work-hardening material, the increase of load required for continuing deformation is regulated both by geometry changes and by the work-hardening of the elements. The eventual collapse occurs through the phenomenon of instability which must be investigated in each case. The value of Q_1 has no application in these systems.

4.2 Geometry changes before uncontained plastic flow

In actual fact, the problem of geometry changes arises even before the stage of uncontained plastic flow. For many problems, if the elastoplastic solution is carried out neglecting the geometry changes as indicated in Section 3.1, uncontained flow occurs only for very large or even infinite deformations. Hence, the changes of geometry leading up to the stage of uncontained flow at a loading value of Q_1 cannot be neglected.

However, from the practical point of view this load is of interest for those problems where Q_1 is 'attained quickly' that means that the value of Q is already very close to Q_1 (for example $0.95 Q_1$) when the deformations are still two to five times as large as those obtained at the initial elastic limit of the system and therefore still negligible.[2] It seems likely that if the accurate solution were carried out, i.e. the geometry changes were taken into account, the results would become noticeably different from those for the case of fixed geometry only at values of Q close to Q_1; and consequently, the load for which important deformation occurs would be close to that given by Q_1. The problem of elasto-plastic flexure of a beam with rectangular cross section subject to uniform bending moment (see [2] and [6]) gives a simple example of this case.

It is not always true that Q_1 is attained quickly, as in the case of a spherical envelope consisting of a perfectly plastic Tresca material, subject to an internal pressure. Here it has been found, by means of an elasto-plastic analysis neglecting geometry changes, that the deformation corresponding to a fixed percentage of Q_1, referred to that obtained at the initial elastic limit, increases considerably as the envelope is made thicker. Thus, it is not possible to give Q_1 a significant meaning with regards to a thick envelope.

The accurate elasto-plastic solution of this problem [6] gives Q_0 as the initial elastic limit of the system. Then, in the elasto-plastic phase, it is found that the load reaches a maximum for Q equal to Q_c called the critical load in [10] and [14]. If the initial thickness of the sphere under investigation is decreased *to arrive at the problem of the thin spherical envelope*, the geometry changes become negligible, and Q_c tends to Q_1 as defined above.

---

[1] Some problems exist in which important geometry changes can be admitted; they produce a considerable 'geometrical work-hardening' and the collapse load is very different from that given by Q_1.
[2] On the assumption that deformations obtained at the elastic limit of the structure are negligible themselves.

Generally speaking, the accurate elasto-plastic solution of each problem is required (i.e. the geometry changes need to be taken into account) to decide whether the limit load Q_1 as defined in Section 3.1 has a practical significance for a perfectly plastic–elastic system. In actual fact, this complex study could be carried out only in exceptional circumstances. It can be seen, by analogy with the problem of the spherical envelope under increasing internal or external pressure, that either one or other of the following cases are attained:

(1). The load increases and reaches a maximum as the plastic deformation develops.

(2). The load increases continuously as the plastic deformation develops.

In the first case the collapse of the structure is defined either by the critical load parameter Q_c, which corresponds to plastic instability, or by a criterion of maximum admissible plastic deformation. In the second case, a criterion of maximum deformation is used. When the geometry changes are negligible, all these criteria will lead to the value Q_1. Thus, the problem lies in estimating a priori whether it is possible to forgo the accurate elasto-plastic investigation and to deal with the collapse of the structure using the limit load Q_1.

The efficiency of the methods available for the determination of Q_1, dealt with in Chapter V, compared to the difficulties of the accurate or even simplified elasto-plastic solution, is the main reason for the importance of the theory of limit loads and of the use of the rigid plastic scheme associated with it. It is essential, however, to be aware of the limits of the application of this theory.

5. Examples

5.1 A simple structure

To illustrate the concepts of elastic limit and yield limit, an example is taken from [5] and [7], as represented in Figure II.3.

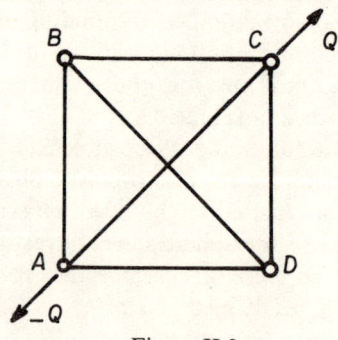

Figure II.3

The system under investigation is a square frame with apices A, B, C, D as hinges. The bars are elastic perfectly plastic and all have the same cross section and properties, the plastic limit having the magnitude L in both tension and compression. The structure is subject to the action of two forces $+Q$ and $-Q$ at A and C. Such loading is dependent upon only one parameter. The system has one degree of redundancy and is assumed to be initially unstressed.

If T denotes the tension in bar AC, we can obtain through statics the forces in all the bars:

$$AC: T, CD \text{ and } CB: (Q - T)\sqrt{2}/2, BD: -(Q - T).$$

For elastic behaviour

$$T = Q\frac{\sqrt{2}}{2}$$

Thus, the greatest force is exerted in AC, and the elastic limit of the structure is attained when this force attains the plastic limit of a bar; hence,

$$Q_0 = L\sqrt{2}$$

If Q exceeds Q_0, the force in AC remains equal to L, i.e. $T = L$. The forces in the other bars are therefore known:

$$CD \text{ and } CB: (Q - L)\frac{\sqrt{2}}{2}, BD: -(Q - L);$$

the deformations in these bars are purely elastic and determinate.

In AC, the deformation is the sum of the elastic and plastic extension. It is the arbitrary value of the plastic part which allows the compatibility of deformation. The plastic deformations are contained since the deformation of the whole structure is that of the elastic structure composed of the five bars AB, BC, CD, DA and BD, subject to forces $(Q - L)$ and $-(Q - L)$. This structure has a degree of redundancy one lower than the initial structure: it is isostatic.

The preceding calculations are valid only until the plastic limit is reached in another bar. BD is the next bar to become plastic, when $Q = 2L$. Now there is uncontained flow: the lengthening of AC and the shortening of BD are not limited by the deformations of the remaining structure as it has now become a mechanism. If the changes of geometry (variation of angles) are still neglected, the flow will continue under a constant load $Q_1 = 2L$, which is the *limit load*.

It is important to note that:

(1). The absence of work-hardening is an essential condition: the force in a bar in limit equilibrium remains constant in spite of any increment of deformation;

(2). The geometry changes are neglected up to the appearance of the uncontained plastic flow. This is possible as the elastic deformations of AV, CD and BD remain small if the Young's modulus of the material is sufficiently large (with respect to L).

5.2 Cavity in an infinite elastic plastic medium

Two problems of elasto-plasticity that are of interest in soil mechanics have been investigated and solved analytically as a result of the symmetry involved.

(1). The expansion of a spherical or cylindrical cavity subject to an internal pressure, in an elasto-plastic infinite medium. The case of a standard Tresca material with or without work-hardening was dealt with in [6]; that of a non-hardening Coulomb material was investigated in [12] for an associated or non-associated flow rule.

(2). The contraction of a cavity under the same conditions [13]. It is essential to allow for geometry changes in these solutions (since a very thick spherical or cylindrical envelope is dealt with).

A finite limit to the pressure is approached asymptotically for the case of expansion, whereas the inner radius of the cavity and the radius of the boundary between the plastic and elastic zones (a ring around the cavity) tend to infinity.

For the case of contraction, the complete crushing of the cavity requires a negative pressure (tension), infinite for the case of a Tresca material, and equal to $-H = -C \cot \phi$ for the case of a Coulomb material. (See, for example, [3] and [14] for further details.)

The solution of this hypothetical problem has numerous applications in soil mechanics, e.g. the equilibrium of very deep subways and subterranean cavities, [1] and [16].

References

[1] F. Baguelin *et al.* (1972) Expansion of cylindrical probes in cohesive soils, *Jl. Soil Mech. & Found. Div., A.S.C.E.*, **98**, No. SMI, Nov. 1972 pp. 1129–1142.
[2] J. Courbon (1965) *Compléments de Résistance des Matériaux. Plasticité*, E.N.P.C., Paris.
[3] P. Habib (1973) *Précis de Géotechnique*, Dunod, Paris.
[4] S. Kobayashi and C. H. Lee (1970) Elastoplastic analysis of plane strain and axy-symmetric flat punch indentation, *Int. J. Mech. Sc.*, **12**, pp. 349–370.
[5] J. Mandel (1964) Contribution théorique à l'étude de l'écroussage et des lois de l'écoulement plastique, *Proc. 11th Congr. Int. Mech. Appl.*, Münich, pp. 502–509.
[6] J. Mandel (1966) *Cours de Mécanique des Milieux Continus*, Gauthier-Villars, Paris.
[7] J. Mandel (1969) *Cours de Science des Matériaux*, Ecole Nationale Supérieure des Mines de Paris.
[8] P. V. Marcal and I. P. King (1967) Elastic-plastic analysis of two-dimensional stress systems by the finite element method, *Int. Jl. Mech. Sc.*, **9**, pp. 143–155.
[9] Q. S. Nguyen and J. Zarka (1972) Quelques méthodes de Résolution Numérique en Plasticité Classique et Viscoplasticité, *Plasticité et Viscoplasticité*, Ed. D. Radenkovic and J. Salençon, Ediscience, Paris, 1974, pp. 327–357.
[10] D. Radenkovic and J. Salençon (1971) *Equilibre limite et rupture en Mécanique des Sols*, Le Comportenent des Sols avant la Rupture, Journées Françaises du C.F.M.S., n° Spéc. Bulletin de Liaison des Laboratoires des Ponts et Chaussées, pp. 296–302.
[11] J. Salençon (1969) La théorie des charges limites dans la résolution des problèmes de plasticite en déformation plate, Thesis Dr. Sc., Paris.
[12] J. Salençon (1966) Expansion d'une cavité ... dans un milieu élastoplastique, *Ann. Pts. Ch.*, 1966, 3, pp. 175–187.
[13] J. Salençon (1969) Contraction d'une cavité ... dans un milieu élastoplastique, *Ann. Pts. Ch.*, 1969, **4**, pp. 231–236.
[14] J. Salençon (1974) Plasticité pour la Mécanique des Sols. C.I.S.M., Rankine Session, July 1974, Udine, Italy.

[15] P. M. Stremsdoerfer (1973) Les méthodes incrémentales en Elastoplasticité-2-Application à l'étude des cavités de stockage de gaz en couche de sel. *R.F.M.*, No. 47, pp. 29–40.
[16] A. S. Vesic (1972) Expansion of cavities in an infinite soil mass, *Jl. Soil Mech. & Found. Div., A.S.C.E.*, **98**, No. SM3, March 1972, pp. 265–290.
[17] O. C. Zienkiewicz, C. Humpheson, and R. W. Lewis (1975) Associated and non-associated viscoplasticity and plasticity in soil mechanics, *Géotechnique*, **25**, No. 4, pp. 671–689.

CHAPTER III

The problems of uncontained plastic flow and an investigation of the 'rigid-plastic model'

1. Definition of a Rigid-plastic Material

A rigid-plastic material is defined as a material in which the only deformation occurring is plastic. The constitutive equation is, therefore,

$$v_{ij} = v_{ij}^{p} \tag{1}$$

It is plain that such a formulation of the behaviour may be related to real problems dealing with elasto-plastic materials only in the limiting case when phenomena are studied for which either the elastic part of the deformation is negligible compared to the plastic part, or the elastic properties have no influence whatever.

2. Statement of the Problem for a Rigid-plastic System

On considering a rigid-plastic material system, the behaviour of which is followed during the loading process, the following remarks may be made *a priori*.

(1). The rigid-plastic pattern will, as a rule, lead to indeterminacies of various kinds; for example, stress fields cannot be wholly determined, in consequence of the non-deformability of the non-plastic regions.

(2). The rigid-plastic pattern allows solution only of problems of uncontained plastic deformation, as non-zero contained plastic deformation cannot occur in a rigid-plastic system. This will have an important influence on the way the problem must be set (see Section 5). Thus, when dealing with a rigid-plastic system it is normal to study first the problems of incipient plastic flow and then, following the changes in geometry step by step, the continuation of uncontained plastic flow.

The unknowns in the problem are no longer the *stress-rates* and velocities, as they were in Chapter II for elasto-plastic problems, but rather the *stresses* and velocities. It is shown in [5] that such a problem is well-defined. The difference of viewpoint between the rigid-plastic and elasto-plastic problems may be explained as follows. For the rigid-plastic system the step by step solution is carried out from the onset of uncontained flow. Thus, the velocities (since the actual displacements are known to be zero) and the stresses must be determined at this moment. The same procedure is repeated at the end of a time-step, in which the displacements may then be determined and are henceforth known.

As already mentioned, the geometry changes must be taken into account during the solution, which is generally difficult. Thence it follows that attention is paid essentially to the problems of:

(1). Incipient uncontained flow, where the geometry is the initial geometry,

(2). Permanent uncontained flow, where the geometry remains unchanged, i.e. a perpetually incipient flow,

(3). Self-similar uncontained flow, where the form of the geometry remains similar to the original.

In actual fact, these three types of problems are of the same nature and correspond, in one way or another, to the study of incipient uncontained flow. The system is then said to be in a *limit equilibrium state*.

3. The Rigid-plastic Pattern and the Determination of Limit Loading

3.1 An example

On studying again the example of the structure dealt with in the previous chapter, it may be seen that the following data participate in the determination of the elastic limit ($Q_0 = L\sqrt{2}$):

(1). *The initial stresses*, which in this example are zero.

(2). *The elastic properties of the material* (though their effect does not appear explicitly in the result, as the assumption that these properties are the same for all bars leads to simplifications).

In the determination of the limit load of the structure, neither the initial stresses nor the elastic properties appear. Effectively, for $T = L$ in AC and $-L$ in BD the value $Q = 2L$ is obtained simply by statics. Member CB has a stress equal to $L\sqrt{2}/2$ and is therefore safe and the structure, which was isostatic for $Q < Q_1$, becomes a *mechanism* for $Q = Q_1$.

Thus, it might seem that, according to this example, the limit load of the system is independent of the elastic properties, in this case the Young's modulus, E, of the constitutive material; and that, in particular, E can be supposed infinite; i.e. it is possible to work on the rigid-plastic system. It must not be forgotten, however, that the study was carried out under the assumption that the geometry changes could be neglected.

3.2 General case; recourse to the rigid-plastic material for the determination of the limit loadings

A system (Σ_1) is constituted by an elastic perfectly plastic material (M_1) of which the elastic moduli at each point M are given by

$$\lambda^1_{ij,hk}(M)$$

A second system (Σ_2) has the same geometry but is constituted by an elastic–perfectly plastic material (M_2) of which the elastic moduli are

$$\lambda^2_{ij,hk}(M) = \rho \lambda^1_{ij,hk}(M) \qquad (\rho > 0)$$

The yield criterion and flow-rule are the same for the two materials.

For the system (Σ_1) a history of loading (H_1) is given by the values of the body forces F^1 and the boundary data $(T_i^d)^1$ on S_{Ti} and (u_i^d) on S_{ui} at moments in time, t.

The loading history (H_2) is given for system (Σ_2) at the same moments, t, by the data

$$F^2 = F^1$$

$$(T_i^d)^2 = (T_i^d)^1 \quad \text{on} \quad S_{Ti}$$

$$(u_i^d)^2 = (u_i^d)^1/\rho \quad \text{on} \quad S_{ui}$$

Then, if the geometry changes may be neglected for (Σ_1) following (H_1), and for (Σ_2), following (H_2), it is obvious that, at each moment t, the stress fields are identical in (Σ_1) and (Σ_2)

$$\sigma^1 = \sigma^2$$

and the velocity fields u^1 and u^2 are related by

$$u^2 = u^1/\rho$$

In particular, in the plastic zones the values of parameter $\lambda \geq 0$, left arbitrary in the flow rule, are λ^1 and λ^2: $\lambda^2 = \lambda^1/\rho$.

The plastic zones are the same in (Σ_1) and (Σ_2) at each moment. Both systems will, therefore, reach the uncontained flow state at the same moment, i.e. for the same limit loading. If, in particular, ρ tends to infinity, it can be seen that system (Σ_∞), following history (H_∞), and constituted by the rigid-plastic material (M_∞) defined by ρ approaching the limit. In system (Σ_∞) the imposed velocities are everywhere zero except on the point of uncontained flow.

Thus, the following result can be stated: for the assumption of negligible geometry changes, the limit loadings of an elastic–perfectly plastic system can be determined by progressing to the limit in the corresponding rigid–perfectly plastic limit system.

Similarly to what was stated in Sections 3.1 and 3.2 the *limit load* (and *limit loading* in the case of several parameters) may be defined as the load corresponding to the appearance of uncontained plastic flow in the corresponding rigid–perfectly plastic limit system.

This rigid–perfectly plastic limit system, for which, moreover, no indeterminacy appears in its resolution, and the rigid–perfectly plastic system (Σ_Ω) defined by [1] are not identical. However, the solution (stress field, velocity field) corresponding to the incipient uncontained plastic flow for (Σ_∞) is also valid for (Σ_Ω). In other words, a limit loading for (Σ_1) and (Σ_∞) is also a limit loading for (Σ_Ω).

The theorem of the *uniqueness of the limit loadings*, later stated in Chapter V, will show the converse of the above property under the hypothesis of the *principle of maximum plastic work*. Thus, from the viewpoint of the limit loadings, it will be proven that it is possible to work directly upon the (Σ_Ω) system independently of any progression to the limit.[1]

4. Governing equations

As stated in Section 3, the problem of a rigid–perfectly plastic system of the type (Σ_Ω) is reduced to the study of the incipient uncontained plastic flow and to the determination of the corresponding limit loadings. With regard to the governing equations, two types of regions may be considered:

(1). *The regions in limit equilibrium*, in which the yield criterion is satisfied at each point;

$$f(\sigma_{ij}) = 0 \qquad (2) \text{ yield criterion}$$

$$\frac{\partial \sigma_{ij}}{\partial x_j} + \rho X_i = 0 \qquad (3) \text{ equilibrium equations}$$

$$V_{ij} = V_{ij}^p \qquad (4) \text{ flow rule}$$

(for instance $V_{ij} = \lambda(\partial f/\partial \sigma_{ij})$, $\lambda \geq 0$ in the case of a standard material).

(2). *The rigid region* in which there is no deformation occurring;

$$V_{ij} = 0 \qquad (5)$$

$$f(\sigma_{ij}) \leq 0 \qquad (6)$$

$$\frac{\partial \sigma_{ij}}{\partial x_j} + \rho X_i = 0 \qquad (3)$$

It can be seen that in the plastic zones exactly sufficient equations to determine the unknowns σ_{ij}, u_i (stresses and velocities) are available. However, in the rigid zones three further equations are necessary to determine the stress field,[2] whilst the velocity field corresponds to motion without deformation (rigid-body motion).

The plastic zones (*a*) are the regions of the system which can be deformed and the rigid zones (*b*) are the undeformed regions. The distinction between these two types of regions may not appear logical since it is possible to belong simultaneously to both types. As a matter of fact it comes from certain peculiarities

---

[1] It will be noticed that in the example of Section 3.1 uniqueness was implicitly taken into account.
[2] This is not surprising since in these regions the stresses do not intervene in the constitutive equation.

of the problems of uncontained flow for a rigid-plastic material, which will be outlined now.

The problem of uncontained flow for a rigid-plastic material is a free boundary problem since the boundary between the plastic zones (*a*) and the non-plastic zones (non-*a*) is not known beforehand, as well as the boundary between the deformed zones (non-*b*) and the undeformed zones (*b*). The particular feature of the problem is that there is no uniqueness of these boundaries: for the same problem different solutions of uncontained plastic flow may happen to be evidenced where the boundaries between the plastic and non-plastic zones are different as well as the boundaries between the deformed and undeformed regions.[1] The only result of uniqueness to be available from that point of view, will be given in Section 6.2 and states that once a point of the system lies in a (non-*b*) zone in a solution for the problem it will belong to (*a*) type zones in any solution.

It follows that the method of solutions of problems of uncontained flow for a rigid-plastic material is rather intricate. As will be seen in detail in Chapter IV (for the particular case of plane-strain problems) the procedure is as follows.

'Solutions' are found, after assuming *a priori* the deformable zones of the system, (*a*), the remainder of the system being undeformed, (*b*). Then, using appropriate theorems (given in Chapter V), the interpretation of these 'solutions' is sought and the actual solutions of the problem can be characterized amongst them.

5. Boundary Conditions

5.1. Classical presentation

As stated previously, with the rigid-plastic scheme it is possible to deal only with uncontained flow problems. The boundary conditions must take this fact into account; i.e. they must be consistent with uncontained flow. This leads to the introduction of consistent boundary conditions, as introduced by Mandel [6, 7]. However, as defined in this text, the problem always reduces to one of determining the limit load, and may be stated as follows. *If the system is subjected to an action of a given type, then the value of this action for which uncontained flow takes place is the solution sought.*

Generally speaking, the boundary conditions consist of data relating to the stresses (surface tractions) and velocities at each point on the boundary of the solid. Three mutually-orthogonal components are prescribed for each of two vectors—**T** for surface traction, and **u** for velocity—as is classical.

However, due to the particular form of the problem, these data must fulfil certain requirements. It is clear, for example, that the three components of the surface tractions could not be prescribed everywhere. The dynamic conditions

---

[1] One can take as an example the problem of the indentation of a rigid-plastic Tresca half plane by a smooth rigid punch, with the two well-known solutions by Shield [10] and Bishop [1], and the velocity fields given Prandtl and Hill.

(i.e. those concerning the forces applied to the system) must be of a somewhat variable character. Thus, the boundary conditions will necessarily contain some data concerning velocity, which in the expression for the work done by the external forces provide the variable dynamic conditions. With the problem being defined with boundary data for the velocity, the solution will necessarily be one of uncontained plastic flow, and the limit loading will be obtained as the corresponding value of the loading.

An example is shown in Figure III.1, involving the indentation of a rigid–perfectly plastic half-plane by a rigid punch moving vertically.

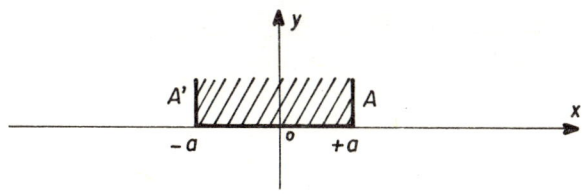

Figure III.1

The boundary conditions are as follows: At ∞,

$$u_x = u_y = 0$$

and at the surface ($y = 0$)

$$|x| > a, \qquad \sigma_y = \tau_{xy} = 0$$
$$|x| < a, \qquad \tau_{xy} = 0, \qquad u_y = u$$

The variable dynamic conditions are, therefore, the distribution of normal stresses along AA'.

5.2 Discussion of the classical presentation

The presentation of the boundary conditions whereby, at each point, three mutually orthogonal components for the set of two vectors **T** and **u** are prescribed is well-known in Continuum Mechanics (elasticity, visco-elasticity, elasto-plasticity, etc.) It has also been adopted by numerous authors for problems concerning a rigid-plastic material. However, it is not always well suited to the case, since in practical problems the dynamic conditions present their variable character in such a way that it is hardly possible to distinguish between the stress and velocity data.

Another example of a plane problem, similar to that of Section 5.1., is the case of a uniform pressure of variable intensity $\bar{\omega}$ applied to the surface of a half-plane (on the segment $A'A$ in Figure III.2). The limit value of $\bar{\omega}$ is required. The condition along $A'A$ in the direction $0y$ cannot be considered as stress or velocity data since the value of $\bar{\omega}$ is not known.

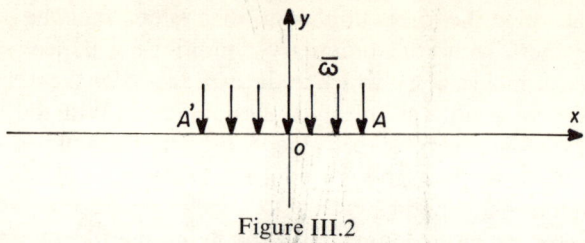

Figure III.2

Similarly, the problem of the indentation of a half-plane by a smooth rigid punch under the action of an axial vertical force F is another example (Figure 3). The condition of $A'A$ in the direction $0y$ is neither stress nor velocity data. It is known that the distribution of u_y must be that of a rigid motion, and the distribution of σ_y corresponds to an axial resultant force.

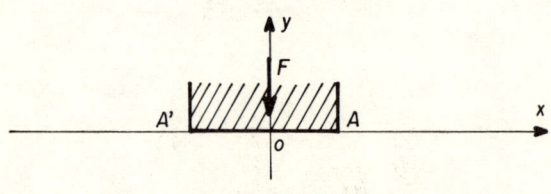

Figure III.3

It is obvious that without invoking artificial transformations, which would affect the clarity, it is not possible to insert these boundary conditions within the standard framework of Section 5.1.

Finally, it should be noted that all the boundary conditions in these practical problems are easily expressed by means of a finite number of loading parameters.[1] For example, in the case of Figure III.2, the boundary conditions depend only on the parameter $Q_1 = \bar{\omega}$; in the case of Figure 3, F is a loading parameter.

Given later as an appendix is a complete presentation of the concept of loading parameters or generalized forces.[2]

A survey of this presentation is proposed now.

The boundary conditions are composed of dynamic and kinematic components that allow the required variable character to the dynamic data.

For any statically admissible stress field, σ, satisfying the dynamic boundary conditions, and any kinematically admissible strain-rate field, v, satisfying the kinematic boundary conditions,[3] the theorem of virtual work yields the equation

---

[1] Because of this there is no insistence on the minimum principles for the stresses and velocities in which the boundary conditions must be written under the form of Section 5.1. Moreover, even when this is possible these 'principles', in spite of their apparent generality are of practical use only in the case of a finite number of loading parameters when they are reduced to the theorems of the theory of limit analysis, as remarked by Hill [3].

[2] Note that in [2] the Q_i denote generalized stresses, i.e. the loading parameters of one element of the system.

[3] When this is necessary, particularly in Chapter V for the exposition of the theory of limit loads, a distinction is made between the local values, labelled **σ**, **v**, **u**, and the corresponding fields labelled σ, v, u.

$$\int_V (\boldsymbol{\sigma} \cdot \mathbf{v}) dV = \int_S \mathbf{T} \cdot \mathbf{u} \, dS + \int_V \rho F \cdot \mathbf{u} \, dV \tag{7}$$

The classical formulation assumes that the reduction

$$\int_S \mathbf{T} \cdot \mathbf{u} \, dS = \int_{S_T} T_i^d u_i \, dS + \int_{S_u} T_i u_i^d \, dS \tag{8}$$

can be effected (T_i^d and u_i^d denote the given values of the components of the stress or velocity on S), which is not always true. The formulation in terms of loading parameters assumes that

$$\int_S \mathbf{T} \cdot \mathbf{u} \, dS = \int_V \rho \mathbf{F} \mathbf{u} \, dV = \sum_{i=1}^n Q_i(\sigma) \dot{q}_i(v) \tag{9}$$

with the following correspondences *being linear*.

$$\sigma \to \mathbf{Q}(\sigma) \in R^n$$
$$v \to \dot{\mathbf{q}}(v) \in R^n$$

This is always possible in the practical cases where the components $Q_i(\sigma)$ of $\mathbf{Q}(\sigma)$ are loading parameters of the system. The vector $\mathbf{q}(v)$ may be termed the strain rate of the system.[1]

The problem therefore resides in determining the limit values of vector $\mathbf{Q}$, together with the corresponding stress and velocity fields.

6. Theorem of Uniqueness of the Stress-field

A theorem of uniqueness for the solution of the problem of uncontained flow will now be given which is valid on the assumption of a *standard* material and a convex yield function (i.e. obeying the principle of maximum plastic work).

6.1 Fundamental inequality

The proofs are founded upon the inequality of the principle of maximum work (Chapter I, Section 6):

$$(\sigma_{ij} - \sigma_{ij}^*) v_{ij} \geq 0 \tag{10}$$

from which

$$(\boldsymbol{\sigma} - \boldsymbol{\sigma}^*) \cdot \mathbf{v} \geq 0$$

---

[1] If the velocity field corresponds to a rigid body motion, $\dot{\mathbf{q}}(v) = 0$.

This inequality implies the convexity of the yield function which is also the plastic potential.

The inequality (10) is not rigidly binding and it is important to investigate under what circumstances the following equality holds true.

$$(\sigma - \sigma^*) \cdot v = 0 \qquad (11)$$

For equality, when $v \neq 0$, both σ and σ^* must belong within or on the yield surface. Since σ is situated on the surface in accordance with the assumption that $v \neq 0$, σ^* cannot be internal since the surface is convex and v is directed along an outward normal at point σ. Therefore, owing to the convexity, the segment $\sigma^*\sigma$ lies wholly within or on the yield surface. On the contrary, according to equation (11), the segment belongs to a place either external or tangential to the yield surface. Consequently for (11) to be true for $v \neq 0$, the segment $\sigma^*\sigma$ must be part of the yield surface. It follows obviously that if this surface is strictly convex, $\sigma = \sigma^*$.

6.2 Theorem of uniqueness

Hill's classical theorem of uniqueness of the stress field is established for the assumption of boundary conditions of the type described in Section 5.1, for which the reduction given in equation (8) is possible. It will be seen later that an analogous theorem is valid for the case of boundary conditions expressed in terms of loading parameters.

6.2.1 Classical statement

For consistent boundary conditions of the type described in Section 5.1 there can exist several solutions, but there is uniqueness of the stress field in the regions formed by combining the deformed zones of these various solutions (for certain complementary assumptions).

6.2.2 Statement of the case for n-parameter loading

Several solutions can exist for boundary conditions of the type given in Section 5.2, corresponding to loading depending on a finite number of parameters Q_i with $\mathbf{Q}$ being the loading for which uncontained flow takes place and $\dot{\mathbf{q}}$ the strain rate of the system at uncontained flow. However, there is uniqueness of the stress field in the regions formed by combining the deformed zones of these various solutions (for certain complementary assumptions).

6.2.3 Proof

For the case of each of the previous statements, two possible solutions are (σ^1, v^1), (σ^2, v^2). Applying the principle of virtual work to fields $(\sigma^1 - \sigma^2)$ and $(v^1 - v^2)$ yields, in the case of Section 6.2.1,

$$\int_V (\boldsymbol{\sigma}^1 - \boldsymbol{\sigma}^2)(\mathbf{v}^1 - \mathbf{v}^2)\,dV = \int_{S_T} (T_i^d - T_i^d)(u_i^1 - u_i^2)\,dS$$

$$+ \int_{S_u} (T_i^1 - T_i^2)(u_i^d - u_i^d)\,dS$$

$$+ \int_V \rho(F - F)(u^1 - u^2)\,dV = 0$$

and in the case of Section 6.2.2,

$$\int_V (\boldsymbol{\sigma}^1 - \boldsymbol{\sigma}^2)(\mathbf{v}^1 - \mathbf{v}^2)\,dV = \sum_{i=1}^{n} (Q_i^1 - Q_i^2)(\dot{q}_i^1 - \dot{q}_i^2)$$

$$= (\mathbf{Q}^1 - \mathbf{Q}^2)(\dot{\mathbf{q}}^1 - \dot{\mathbf{q}}^2) = 0$$

since it is assumed that $\mathbf{Q}^1 = \mathbf{Q}^2$ (resp. $\dot{q}^1 = \dot{q}^2$).

The remainder of the proof is then similar in both cases.

It has been shown that

$$\int_V (\boldsymbol{\sigma}^1 - \boldsymbol{\sigma}^2)(\mathbf{v}^1 - \mathbf{v}^2)\,dV = 0 \tag{12}$$

whence, applying equation (10),

$$(\boldsymbol{\sigma}^1 - \boldsymbol{\sigma}^2)(\mathbf{v}^1 - \mathbf{v}^2) = 0 \tag{13}$$

at any point of the system (V). The left-hand side of equation (13) is zero in the following cases:

(1). If $\mathbf{v}^1 = \mathbf{v}^2 = 0$ (i.e. in the zones which are rigid in both solutions),

(2). In the zones that are deformed in at least one solution (i.e. in the regions formed by the combination of the deformed zones in each solution).

If only one tensor $\mathbf{v}$ is non-zero, e.g. $\mathbf{v}^1$, then equation (13) becomes

$$(\boldsymbol{\sigma}^1 - \boldsymbol{\sigma}^2)\mathbf{v}^1 = 0$$

which is equation (11), studied in Section 6.1. If both tensors are non-zero, equations (13) and (11) imply that

$$(\boldsymbol{\sigma}^1 - \boldsymbol{\sigma}^2)\mathbf{v}^1 = 0 \quad \text{and} \quad (\boldsymbol{\sigma}^2 - \boldsymbol{\sigma}^1)\mathbf{v}^2 = 0$$

again equation (11). From the conclusions of Section 6.1 we can deduce that $\boldsymbol{\sigma}^1$ and $\boldsymbol{\sigma}^2$ belong to the yield surface:

$$f(\boldsymbol{\sigma}^1) = f(\boldsymbol{\sigma}^2) = 0$$

In addition the segment $\boldsymbol{\sigma}^1\boldsymbol{\sigma}^2$ belongs to the yield surface, from which it follows that if this surface is regular (i.e. there is only one tangent plane at each point), then $\mathbf{v}^1$ and $\mathbf{v}^2$ are colinear.

Thus, if we define D as being the combination of the deformed zones corresponding to each solution, the following conclusions can be made.

(1). In any solution, D is in a limit equilibrium state.

(2). If the yield surface is regular, there is at each point of D a proportionality between the strain rates of the different solutions (in particular, the principal directions are the same). In the case of isotropy, this implies the coincidence of the principal directions of the stresses.

(3). If f is strictly a convex function of σ, the stress field is determined in a unique way at any point of D.

(4). If $\bar{f}$ is strictly a convex function of the deviator $\mathbf{s}$ (Mises' criterion for example), $\mathbf{s}$ is determined in a unique way at each point of D (since $\sigma^1\sigma^2$ is a generatrix of the cylinder). σ is determined in a unique way at each point of D if the normal stress is known in at least one point on the boundary of D.

In the case of the Tresca or Coulomb criterion (2) may be replaced by the following statement.

(2a). The flow regime at any point of D is the same in all solutions (same face or same edge). In the case of the face regime, there is proportionality between the strain rates in the different solutions (in particular, they have the same principal directions).

The statements corresponding to (3) and (4) are more complicated (see [5]).

7. Remarks

On reading the previous paragraphs, a sense of doubt may arise as to the usefulness of the rigid-plastic behaviour pattern. Indeed, the modelling of the real elastic-plastic medium by the rigid-plastic medium can bring many complications, as demonstrated in Section 5. Moreover the question of the exact meaning of the rigid-plastic system (Σ_r) with respect to the elasto-plastic one (Σ_λ) has not been thoroughly dealt with (Section 4). Only will it be settled in Chapter V when the identity of (Σ_r) with any (Σ_∞) will be proved under the assumption of the principle of maximum work. It is our opinion however that the rigid-plastic behaviour pattern has to be associated with the principle of maximum work in one way or another in order to be given a significance.

In spite of difficulties, the rigid-plastic pattern has proved to be very useful for, as we shall see in the following chapters, it made it possible to solve a number of problems engineers had to overcome. Also it must not be forgotten that the adopted formulation should be adapted to the problem under investigation and to the method of computation available. It may be that at some future date either the elasto-plastic, visco-plastic, or the pseudo-plastic material will become the most useful pattern, as a consequence of the development of new computational means.

References

[1] J. F. W. Bishop (1953) On the complete solutions of deformation of a plastic rigid material, *J. Mech. Phys. Solids*, **2**, pp. 43–53.

[2] J. Courbon (1969) *Notions de Plasticité*, Cours E.N.P.C., Paris.
[3] R. Hill (1951) On the state of stress in a plastic rigid body at the yield point, *Phil. Mag.*, **42**, pp. 868–875.
[4] W. T. Koiter (1960) General theorems for elastic plastic solids, In *Progress in Solid Mechanics*, Ed. Sneddon and Hill, North-Holland Publ. Co., Amsterdam.
[5] J. Mandel (1965) Sur l'unicité du champs des contraintes lors de l'équilibre limite dans un milieu rigide-plastique, *C.R.Ac.Sc.*, Paris, **261**, pp. 35–37.
[6] J. Mandel (1966) *Mécanique des Milieux Continus*, Vol. II, Gauthier-Villars, Paris.
[7] J. Mandel (1972) Unicité et principes variationnels en viscoplasticité, *Plasticité et Viscoplasticité*, Eds. D. Radenkovic and J. Salençon, Ediscience, Paris, 1974, pp. 186–202.
[8] J. Salençon (1969) La théorie des charges limites dans la résolution des problèmes de plasticité en déformation plane, Thesis Doct. Sc., Paris.
[9] J. Salençon (1974) *Plasticité pour la Mécanique des Sols*, C.I.S.M., Rankine Session, Udine, Italy, 1974.
[10] R. J. Shield (1954) Plastic potential theory and Prandtl bearing capacity solution, *J. Appl. Mech. trans. A.S.M.E.*, **21**. pp. 193–194.

CHAPTER III

Appendix

GENERAL DEFINITION OF THE LOADING PARAMETERS FOR A SYSTEM

1. Possible Loadings

1.1 Boundary conditions

A deformable body (V) is considered with a volume V and a boundary $\partial V = S$. The boundary data concerning both the stress vector **T** and the velocity **u** consist of

(1). Three mutually orthogonal components for the set of these two vectors at each point of the boundary S;
(2). The body forces throughout V.

Mx_i ($i = 1, 2, 3$) at each point M of S, are the orthogonal axes corresponding to the directions of the given components of **T** and **u**.

Attention is given to the problems of equilibrium of the body (quasi-static deformation) under all the loadings compatible with equilibrium for which the axes Mx_i remain the same ($\forall i$) as well as the nature (**T** or **u**) of the datum along Mx_i.

These loadings will be called possible loadings of (V).

The part of S on which the component T_i is given will be called S_{T_i}; and the part of S on which the component u_i is given will be called S_{u_i}. It is obvious that

$$\forall i \quad S_{T_i} \cap S_{u_i} = \phi \qquad S_{T_i} \cup S_{u_i} = S$$

1.2 Definitions

The suit of possible dynamic data, J_d, has components T_i over S_{T_i} ($i = 1, 2, 3$), **F** (body forces) throughout V (such that equilibrium is possible).

The suit of possible kinematic data, J_c, has components u_i over S_{u_i} ($i = 1, 2, 3$).

A suit of possible data, J, is formed by the union of a suit of possible dynamic data and a suit of possible kinematic data:

$$J = J_d \cup J_c$$

It is easily seen that the sets of all suits of possible dynamic data, of all suits of possible kinematic data, and of all suits of possible data, may be given structures of linear spaces in R. The first two of those spaces, generally of infinite dimension, will be denoted by $\mathscr{D}$ and $\mathscr{C}$.

A *statically admissible stress-field* is a stress-field σ associated with a suit of possible dynamic data, and expressed as

$$\sigma \text{S.A. ass. } J_d \in \mathscr{D},$$

if, for this suit of data, it satisfies

(1). The equations of equilibrium (in the weak sense),
(2). The boundary data for the stresses.

A *kinematically admissible strain-rate field* is a strain-rate field v associated with a possible suit of kinematic data expressed as

$$v \text{K.A. ass. } J_c \in \mathscr{C},$$

if its components

$$v_{ij} = \frac{1}{2}\left(\frac{\partial u_i}{\partial x_j} + \frac{\partial u_j}{\partial x_i}\right)$$

derive (in the weak sense) from a velocity field u satisfying the boundary conditions in this suit of data.

1.3 Theorem of virtual work

It is assumed that

$$\forall \sigma \text{ S.A. ass. } J_d, \quad J_d \in \mathscr{D},$$

$$\forall v \text{ K.A. ass. } J_c, \quad J_c \in \mathscr{C}.$$

Noting that $\sigma \cdot \mathbf{n} = \mathbf{T}$ at each point of S and denoting by u the velocity field from which v is derived,

$$\int_S \mathbf{T}\mathbf{u}\,dS + \int_V \rho \mathbf{F}\mathbf{u}\,dV = \int_V (\sigma \cdot v)\,dV = \mathscr{T}(\sigma, v).[1]$$

where $\mathscr{T}(\sigma, v)$ is bilinear form (functional) of σ and v.

2. Loading Process Depending on a Finite Number of Parameters

2.1 Definitions

Of all the possible loadings of Section 1, consideration is given only to those that are actually feasible within the frame of a particular loading process. The body is then said to be subjected to a loading process that depends on a finite number of parameters Q_i if

---

[1] The notation is $(\boldsymbol{\sigma} \cdot \mathbf{v}) = \sigma_{ij} \cdot v_{ij}$.

(1). The set of all the permitted suits of dynamic data constitute a linear sub-space $\mathscr{D}_p \subset \mathscr{D}$;

(2). The set of all the permitted suits of kinematic data is a linear sub-space $\mathscr{C}_p \subset \mathscr{C}$;

(3). Two linear applications can be determined,

$$\forall J_d \in \mathscr{D}_p, \quad \forall \sigma \text{ S.A. ass. } J_d,$$

where

$$\sigma \to \mathbf{Q}(\sigma) = [Q_i(\sigma), \ldots, Q_n(\sigma)] \in R^n$$

and

$$\forall J_c \in \mathscr{C}_p, \quad \forall v \text{ K.A. ass. } J_c$$

where

$$v \to \dot{\mathbf{q}}(v) = [\dot{q}_i(v), \ldots, \dot{q}_n(v)] \in R^n,$$

both being such that

$$\mathscr{T}(\sigma, v) = \mathbf{Q}(\sigma) \cdot \dot{\mathbf{q}}(v) = \sum_{i=1}^{n} Q_i(\sigma) \dot{q}_i(v).$$

The $Q_i(\sigma)$ values are the loading parameters of the body.

2.2 Properties

The vectors $\mathbf{Q}(\sigma)$ corresponding to all the stress fields σ S.A. associated with all the feasible suits of dynamic data, constitute an n-dimension linear space, $\{\mathbf{Q}\}$. A vector $\mathbf{Q}(\sigma)$ will be called a loading of the body. Likewise, the $\dot{\mathbf{q}}(v)$ vectors corresponding to all the strain-rate fields associated with all the sets of kinematic data constitute an n-dimension linear space $\{\dot{\mathbf{q}}\}$. $\{\mathbf{Q}\}$ and $\{\dot{\mathbf{q}}\}$ are dual.

2.3 Remarks

For a given $J_d \in \mathscr{D}_p$ there are usually several possible stress-fields, and several $\mathbf{Q}(\sigma)$ vectors corresponding to these fields.

Actually, if

$$\sigma_1 \text{ S.A. ass. } J_d \in \mathscr{D}_p$$

and

$$\sigma_2 \text{ S.A. ass. } J_d \in \mathscr{D}_p$$

then

$$\sigma_1 - \sigma_2 \text{ is S.A. ass. } 0 \in \mathscr{D}_p,$$

but it does not necessarily follow that $\mathbf{Q}(\sigma_1) = \mathbf{Q}(\sigma_2)$. An analogous property holds for $J_c \in \mathscr{C}_p, v, \dot{q}(v)$.

A single **Q** vector can correspond to several S.A. fields σ but all of them are associated with a single $J_d \in \mathcal{D}_p$. An analogous property is valid for $\dot{q}$, v (the J_c are then identical apart from rigid-body motion.) A $\dot{q}(v)$ vector is called a strain-rate of a body.[1] It has been assumed in the definition in Section 2.1. that $\mathcal{D}_p$ and $\mathcal{C}_p$ are linear sub-spaces. This implies that, in the loading process, the constant data must be zero. For the case in which constant data would exist (for example, dynamic data representing body forces or load on part of a surface), these should be considered as variable, and only given their prescribed values at the end.

3. Case of a System with Friction Conditions

Statements in Sections 1 and 2 concern the case of a body with boundary data belonging to the type indicated in Section 1.1. The application to the case of a body, or system of bodies, in which friction conditions intervene (non-smooth contacts) requires further consideration. The loading parameters, in the case of body with a friction contact on part of its boundary are defined for a larger system including both the body and the friction interface. (The system's boundary passes outside this interface.) The friction condition only appears as a flow rule within the system. The same applies for a system comprising several bodies coming into contact with friction.

4. Example

Indentation by a smooth rigid plate of a half-plane with a uniformly distributed surface load (Figure III.A.1).

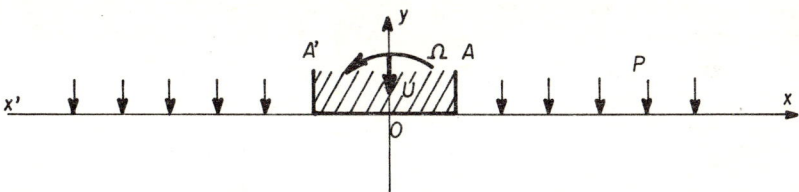

Figure III.A.1

At ∞,

$$\text{for } S_{u_1}, u_1 = 0;$$
$$\text{for } S_{u_2}, u_2 = 0.$$

On $x'A'$ and Ax,

$$\text{for } S_{T_1}, T_1 = 0;$$
$$\text{for } S_{T_2}, T_2 = -p. \quad (p \text{ is arbitrary}).$$

---

[1] This nomenclature makes it apparent that if the velocity field u is that of a rigid-body motion, $\dot{q}(v) = 0$.

On $A'A$

for S_{T_1}, $T_1 = 0$;

for S_{u_2}, $u_2 = -U + \Omega x$. (U, Ω are arbitrary).

Hence,

$$Q_1 = p, \quad \dot{q}_2 = U, \dot{q}_3 = \Omega,$$

and

$$\dot{q}_1 = -\int_{-\infty}^{-a} u_2 \, dx - \int_a^\infty u_2 \, dx, \quad Q_2 = N, \quad Q_3 = M.$$

(N is the resultant of the forces applied by the punch, measured positively in the $-0y$ direction, M is the moment with respect to 0).

CHAPTER IV

Problems of uncontained plastic flow in plane strain

1. General

This chapter deals with a class of uncontained plastic flow problems the solution of which incorporates important simplifying features. For this reason the form of many fundamental Soil Mechanics problems has been reduced to that of plane plastic flow.

These problems will be studied very closely, as they are rather difficult, especially in the logic involved which is sometimes intricate.

The probems of axial symmetry can be solved similarly for certain hypotheses (e.g. the criterion of the intrinsic curve type, Haar–Karman's hypothesis [21]). These problems are dealt with in Appendix B and readers requiring more complete information are advised to read references [1, 12, 21, 39, 54, 60].

Only the case of isotropic materials will be dealt with though anisotropy has been studied in plane plastic strain [2, 7, 13, 25, 26].

2. Expression of the Yield Criterion

The plane strain normal to the $0z$ axis is defined by the following conditions.

$$u_x \text{ and } u_y \text{ are independent of } z; \quad u_x = 0 \tag{1}$$

whence,

$$v_{xz} = v_{yz} = v_{zz} = 0 \tag{2}$$

In the deformed, i.e. plastic zones, in which the factor $\lambda \neq 0$, equation (2) represents three equations which generally make it possible to express σ_z, τ_{xz}, τ_{yz} as functions of σ_x, σ_y, τ_{xy}. The yield criterion satisfied in these regions can, therefore, be written as a function of only three components (relating to the plane (x, y)):

$$F(\sigma_x, \sigma_y, \sigma_z, \tau_{xy}, \tau_{xz}, \tau_{yz}) = f(\sigma_x, \sigma_y, \tau_{xy}) = 0.[1] \tag{3}$$

---

[1] There is no indication of the dependence of F and f on x, y if the material is not homogeneous (homogeneous only along $0z$); the results of Sections 2, 3 and 4 are valid in this case.

Since only the case of isotropic materials is dealt with the yield criterion may be expressed as a function of the principal stresses:

$$F(\sigma_1, \sigma_2, \sigma_3) = 0 \tag{4}$$

Under plane strain conditions, $0z$, being a principal direction of **v**, is also a principal direction for **σ**.[1] Setting $\sigma_z = \sigma_3$ (with unordered principal stresses) allows equation (2) to be written as

$$v_3 = 0 \tag{5}$$

Assuming the material obeys the principle of maximum plastic work it follows from equation (5) that

$$\frac{\partial F}{\partial \sigma_3} = 0 \tag{6}$$

Thus, equations (4) and (6) define the yield criterion in plane plastic strain, viz.

$$\left. \begin{array}{l} F(\sigma_1, \sigma_2, \sigma_3) = 0 \\[2pt] \dfrac{\partial F}{\partial \sigma_3} = 0 \end{array} \right\} \quad \begin{array}{l}(4)\\[2pt](6)\end{array}$$

This criterion, expressed as a function of the two principal stresses σ_1 and σ_2 in the plane (x, y),

$$f(\sigma_1, \sigma_2) = 0 \tag{7}$$

is obviously, as a consequence of the convexity of F, the equation of the projection in the σ_3 direction of the apparent boundary of the surface $F = 0$ onto the plane of σ_1 and σ_2. This is a real curve symmetrical with respect to the bisector ($\sigma_1 = \sigma_2, \sigma_3 = 0$). Thus, f is symmetrical in σ_1 and σ_2, and may be written in the form

$$g\left[\frac{\sigma_1 + \sigma_2}{2}, \frac{|\sigma_1 - \sigma_2|}{2}\right] = 0 \tag{8}$$

i.e. in the plane (x, y) a criterion of the intrinsic curve type exists.

With the notation of Chapter I equation (8) reads:

$$g(-p, R) = 0 \tag{9}$$

where R is the radius of the Mohr circle and $-p$ the abscissa of its centre. Later, the intrinsic curve is assumed to be real.

The theorem derived may be summarized as follows.

---

[1] It is stated in Chapter I that in the case of isotropy **v** and **σ** necessarily have the same principal directions. A very detailed discussion of this result arising from the constitutive law linking **v**, **σ** and $D\sigma/Dt$ (in which the relationship between **v** and $D\sigma/Dt$ is linear) will be found in [25]. (See also [2] and [30]).

For an isotropic, rigid-plastic material obeying the principle of maximum plastic work, any yield criterion in plane plastic strain reduces to a criterion of the intrinsic curve type in the plane of the strain.

In particular, the criteria which depend only on the deviator stress (ductile materials), always result in the two-dimensional Tresca criterion, i.e.

$$|\sigma_1 - \sigma_2| = 2k \tag{10}$$

3. A Remark on the Case of a Non-standard Material

For an isotropic material, with a yield criterion of the intrinsic curve type, e.g. a soil obeying the Coulomb criterion, it may not be necessary to refer to the principle of maximum plastic work to arrive at formula (8) or (9): it is sufficient that σ_z, a principal stress as defined previously, is the *intermediate* principal stress.

This may be a consequence of the assumption of plane plastic strain and a particular flow rule, not necessarily that of a standard material. It occurs, for example, if a flow rule similar to equation (25) proposed in Chapter I is used, or if the constitutive law of a standard Mises material is adopted.

It is also true for an isotropic material with a yield criterion of the general form $F(\sigma_1, \sigma_2, \sigma_3) = 0$ (F being convex and symmetrical), and plastic strain obeying the flow rule of a standard Mises material. Again, in the plane of the strain the criterion is represented by a real curve symmetrical with respect to the bisector ($\sigma_1 = \sigma_2, \sigma_3 = 0$), i.e. a criterion of the intrinsic curve type. From hereon it will always be assumed that the envelope of the Mohr circles is real.

Under the conditions stated the results to be developed for the stress field are valid for such non-standard materials.

4. Equations for the Stresses

Under the specified conditions for the validity of equations (8) or (9), three equations are available for the three unknown stresses in the plastic zones, which are independent of the velocities.

$$\begin{cases} f(\sigma_x, \sigma_y, \tau_{xy}) = 0 & (3) \\ \dfrac{\partial \sigma_x}{\partial x} + \dfrac{\partial \tau_{xy}}{\partial y} + \rho X = 0 & (11) \\ \dfrac{\partial \tau_{xy}}{\partial x} + \dfrac{\partial \sigma_y}{\partial y} + \rho Y = 0 & \end{cases}$$

The main interest in plane plastic strain problems is due to this particular circumstance. If the boundary conditions are appropriate, it is possible to proceed in two stages, starting with the determination of the stress field in the plastic zones, without involvement of the velocities, and only then determining the velocity field.

However, this procedure proves to be more complex than indicated here, and it is often necessary to make use of the velocities for the determination of the stress-field, e.g. to select one out of two possible stress solutions. Nevertheless, this possibility of dealing firstly with the stresses and then with the velocities, does make the work considerably simpler.

It should be noted, moreover, that the system of equations (3, 11, 12) and the accompanying conclusions are normally valid in zones where σ_z is the intermediate principal stress and a state of limit equilibrium prevails, when the material obeys a yield criterion of the intrinsic curve type. Therefore the results of the following Section A can be applied in elasto-plasticity problems to zones in limit equilibrium ($f = 0$) which meet these conditions.

In the rigid zones (denoted as b-type zones in Chapter III), three equations are needed to determine the stress field if it is assumed that σ_z is a principal stress. One equation is lacking in order to determine the three stresses $\sigma_x, \sigma_y, \tau_{xy}$, as the yield criterion supplies no more than an inequality.

A. THE STRESS PROBLEM

5. Transformation of the Equations

The method adopted follows Mandel's approach [35, 36, 38].

The principal stresses σ_1 and σ_2 are ordered in the plane O_{xy} following $\sigma_1 \geq \sigma_2$. A new variable is introduced, the angle $\theta = (Ox, \sigma)$, which is the angle between the direction of the greatest traction and the axis Ox.

The stresses $\sigma_x, \sigma_y, \tau_{xy}$ are expressed as functions of the three variables R, p, θ, by means of Figure IV.1:

$$\left. \begin{array}{l} \sigma_x = -p + R \cos 2\theta \\ \sigma_y = -p - R \cos 2\theta \\ \tau_{xy} = R \sin 2\theta \end{array} \right\} \quad (13)$$

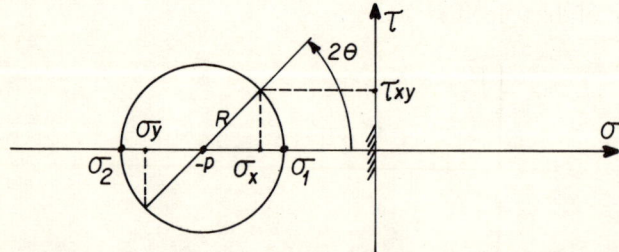

Figure IV.1

From hereon the material is considered as homogeneous, and the case of a material which is non-homogeneous in the plane of the strain (when f is a function of x and y) is studied in an appendix.

The yield criterion (independent of x and y) is satisfied at any point of the plastic zone and is solved in the form

$$R = R(p), \qquad (14)$$

(the solution of equation (9)).
According to relation (13),

$$\frac{\partial \sigma_x}{\partial x} = -\frac{\partial p}{\partial x} + \frac{dR}{dp}\frac{\partial p}{\partial x}\cos 2\theta - 2R \sin 2\theta \frac{\partial \theta}{\partial x} \qquad (15)$$

and likewise for

$$\frac{\partial \tau_{xy}}{\partial x}, \frac{\partial \tau_{xy}}{\partial y} \quad \text{and} \quad \frac{\partial \sigma_y}{\partial y}$$

The derivative dR/dp is easily estimated from Figure IV.2. Here, ϕ denotes the angle between Ox and the tangent to the intrinsic curve at the point of contact with the Mohr circle having a centre of abscissa $-p$.

$$\frac{dR}{dp} = \sin \phi \qquad (16)$$

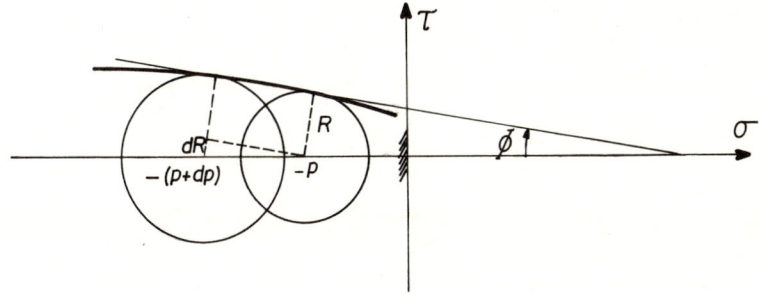

Figure IV.2

The system of equations for the stresses in the plastic zone is thus transformed by the introduction of new variables. Equation (3) is equivalent to (13) and (14), and equations (11) and (12) become

$$-\frac{\partial p}{\partial x}(1 - \sin \phi \cos 2\theta) - 2R \sin 2\theta \frac{\partial \theta}{\partial x} + \sin \phi \frac{\partial p}{\partial y} \sin 2\theta$$

$$+ 2R \cos 2\theta \frac{\partial \theta}{\partial y} + pN = 0 \qquad (17)$$

$$+ \frac{\partial p}{\partial x} \sin \phi \sin 2\theta + 2R \cos 2\theta \frac{\partial \theta}{\partial x} - \frac{\partial p}{\partial y}(1 - \sin \phi \cos 2\theta)$$

$$+ 2R \sin 2\theta \frac{\partial \theta}{\partial y} \div pY = 0 \qquad (18)$$

This is a quasi-linear system of two equations of the first order for the unknown functions, p and θ, of the two variables x and y. It can be shown that if the envelope of the circles defined by relation (14) is real, i.e. if the intrinsic curve is real ($|\sin \phi| \leq 1$), the system of equations (17, 18) is always hyperbolic. Also, there are two families of real characteristic lines.

6. Characteristic Lines

Any classical method of determining the characteristics shows[1] that, at any point of the plastic zone, they are defined by

$$\frac{dy}{dx} = \tan\left[\theta \pm \left(\frac{\pi}{4} + \frac{\phi}{2}\right)\right] \qquad (91)$$

Thus the characteristics make an angle $\pm(\pi/4 + \phi/2)$ with the direction of the greatest traction. By convention, the slip line inclined at $-(\pi/4 + \phi/2)$ to σ_1 is labelled α, and the other β (see Figure IV.3).

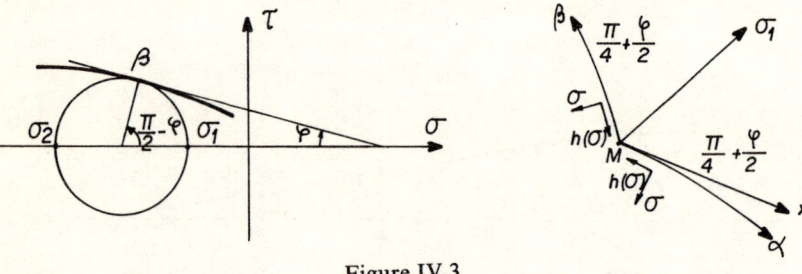

Figure IV.3

At M, the Mohr circle is a tangent to the intrinsic curve, as this point is in a state of limit equilibrium (plastic zone). Equation (19) shows that the characteristic directions at point M correspond to the points of contact of the Mohr circle with the intrinsic curve. These are the two surfaces on which $|\tau| = h(\sigma)$ (see Chapter I, Section 2).

7. Relations Along the Characteristic Lines

It is known that along each characteristic line, the solution (p, θ) of the problem must satisfy a differential relation. This relation states that Cauchy's problem along the characteristic line, with the given values of (p, θ) for the solution, is not impossible but indeterminate.

These relations can easily be established. Assuming that axis Ox follows the tangent to $M\alpha$ at point M, inclined at $-(\pi/4 + \phi/2)$ to σ_1 (Figure IV.3), equations (17) and (18) become

---

[1] For example, writing that, along a characteristic, Cauchy's problem is impossible or indeterminate.

$$-\frac{\partial p}{\partial x}(1 + \sin^2 \phi) - 2R \cos \phi \frac{\partial \theta}{\partial x} + \sin \phi \frac{\partial p}{\partial x} \cos \phi$$

$$- 2R \sin \phi \frac{\partial \theta}{\partial y} = pX = 0 \qquad (20)$$

$$\frac{\partial p}{\partial x} \sin \phi \cos \phi - 2R \sin \phi \frac{\partial \theta}{\partial x} - \cos^2 \phi \frac{\partial p}{\partial y}$$

$$+ 2R \cos \phi \frac{\partial \theta}{\partial y} + \rho Y = 0 \qquad (21)$$

On multiplying (20) by $\cos \phi$ and (21) by $\sin \phi$ and adding, a relation is obtained in which only partial derivatives with respect to x appear:

$$-\frac{\partial p}{\partial x} \cos \phi - 2R \frac{\partial \theta}{\partial x} + \rho X \cos \phi + \rho Y \sin \phi = 0 \qquad (22)$$

This is the differential relation along the characteristic line α,

$$- dp - \frac{2R}{\cos \phi} d\theta + \rho(X + X \tan \phi) dS_\infty = 0 \qquad (23)$$

$X + Y \tan \phi = A$ is the α oblique component of the body force in the α, β axes, whence

$$dp + \frac{2R}{\cos \phi} d\theta = \rho A \, ds_\alpha = 0 \qquad (24)$$

and likewise, along the β characteristic line

$$dp - \frac{2R}{\cos \phi} d\theta = \rho B \, ds_\beta = 0.[1] \qquad (25)$$

If the material obeys the Coulomb criterion, then equations (24, 25) due to Mandel [35], in the case of a homogeneous material and for any intrinsic curve, reduce to Kötter's equations (31, 32).

In the Appendix to Chapter V, which deals with Bonneau's theorem, Kötter's equations are obtained by considerations that clearly indicate their physical meaning.

8. Computation of the Solution

The mathematical properties described in Sections 5, 6, and 7 are of great practical interest, as they make it easy to compute the solution of the stress problem in the plastic zones. As the problem is hyperbolic, the classical method of characteristics (Massau's method [41]) can be used to find the solution.

---

[1] In Appendix A, dealing with non-homogeneous material (and equally applying to homogeneous material), equations (24) and (25) are written with the classical tensorial notation.

The values of functions p and θ along a non-characteristic arc AB are assumed to be known. The continuous solution of equations (17, 18) is uniquely determined on each side of AB in the curvilinear triangle bounded by AB and the inner characteristic lines issuing from A and B. The area is the so-called domain of determinacy of AB. (The proof of this classical result of the theory of partial differential equations is outlined by the method of characteristics).

The method of characteristics allows computation of the solution by discretization. The curve AB is divided by n points $1, 2, 3 \ldots n$ (Figure IV.4). At each of these points, the slopes of the characteristic lines α and β are known from relation (19). The assimilation of these characteristics with their tangents at points $1, 2, \ldots n$ (which becomes more acceptable as the division of AB gets finer) allows the determination of points $(1, 2), (2, 3) \ldots, (n-1, n)$, which approximate to the intersections of the characteristic lines α and β issuing from points $1, 2, 3, \ldots, n$ At these intersection points p and θ may be calculated. At each point $(k, k+1)$ the relations necessarily satisfied by the solution along the characteristics $\alpha[k+1, (k, k+1)]$ and $\beta[k, (k, k+1)]$ supply, by means of finite difference relationships, two linear equations for the two unknowns.

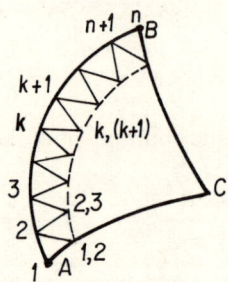

Figure IV.4

The process can then be repeated by making the points $(k, k+1)$ (amounting to a number $n-1$) play the role of the points k in the preceding stage. Thus, the solution is calculated along the characteristics; whence the result already stated concerning the domain of determinancy. Clearly, the finer the mesh, the more accurate will be the approximation of the solution obtained.

This method can also be used to compute the solution from data of p and θ on two converging arcs of characteristics, as well as in other cases (see, for example, [26].)

It is not intended to become involved in the details of numerical analysis but merely to give the principles of the method of characteristics, using the simplest pattern of finite differences. Obviously, more complex patterns can be used[1] both to determine the intersections of the grid and to discretize the relations along the characteristics. The essential point is that the solution is found by using the characteristic lines and the relations along them.

---

[1] Note that both the mesh refinement and the refinement of the discretization pattern contribute to an improvement of the accuracy of a numerical solution.

The method of characteristics is the basic tool for the solutions of plane flow problems and is usually carried out numerically. Only in exceptional cases can analytical solutions be found, as indicated later in a classical example. The method has been much used by Sokolovski [55–58].

9. Transformation of Equations (24, 25)

In the case of no body forces, equations (24) and (25) may be written

$$dp + \frac{2R}{\cos \phi} d\theta = 0 \quad \text{along an } \alpha \text{ line}$$

$$dp - \frac{2R}{\cos \phi} d\theta = 0 \quad \text{along a } \beta \text{ line}$$

In accordance with Mandel, Σ is defined by

$$\Sigma = -\frac{1}{2} \int \frac{\cos \phi(p)}{R(p)} dp \tag{26}$$

where $R = R(p)$ is relation (14), and $\phi = \phi(p)$ is the relation between the abscissa of the centre of a limiting Mohr circle and the angle of its tangent at the point of contact with the intrinsic curve. Then (24) and (25) are integrated to give

$$\Sigma - \theta = \text{const.} = \beta \text{ along an } \alpha \text{ line}$$
$$\Sigma + \theta = \text{const.} - \alpha \text{ along an } \beta \text{ line} \tag{27}$$

For Tresca's criterion

$$\Sigma = -P/2k$$

which results in Hencky's relations [24]:

$$p + 2k\theta = \text{const. along an } \alpha \text{ line};$$
$$p - 2k\theta = \text{const. along a } \beta \text{ line.} \tag{28}$$

For Coulomb's criterion,

$$|\tau| = C - \sigma \tan \phi,$$
$$R = (H + p) \sin \phi \quad \text{where} \quad H = C \cot \phi$$

On the other hand,

$$dp = dR/\sin \phi \tag{16}$$

which yields

$$\Sigma = -\frac{\cot \phi}{2} \int \frac{dR}{R} = -\frac{\cot \phi}{R} \log R$$

or

$$\Sigma = -\frac{\cot \phi}{2} \log (H + p) \tag{29}$$

9.2 Case of a Tresca criterion with conservative body forces

When dealing with Tresca's criterion, $\phi = 0$ and the characteristic lines constitute a net of *orthogonal curves*.

If the body forces are derived from a potential V, as is most common, then

$$A = -\frac{\partial V}{\partial s_\alpha}, \quad B = -\frac{\partial V}{\partial s_\beta}$$

Equations (24) and (24) then become,

$$d(p + \rho V) + 2k\, d\theta = 0, \text{ for an } \alpha \text{ line}$$
$$d(p + \rho V) - 2k\, d\theta = 0, \text{ for a } \beta \text{ line}$$
(30)

from which are derived equations (27) and (28), modified by the replacement of p by $(p + \rho V)$.

10. Geometry of the Characteristic Network

10.1

The characteristic lines constitute a network of two families of curves that intersect at an angle $(\alpha, \beta) = (\pi/2 + \phi)$.

10.2 Hencky's theorem

For the above cases, where the relations along the characteristics are integrated in the form of equation (27) (excluding the case of a Coulomb soil with self-weight), the intersection of pairs of α and β lines (at M, N, P, Q in Figure IV.5) is now considered.

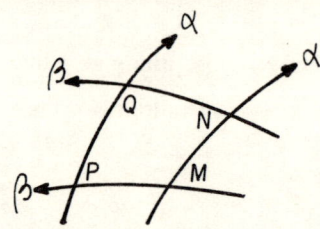

Figure IV.5

From equations (27),

$$\Sigma_M - \theta_M = \Sigma_N - \theta_N$$
$$\Sigma_P - \theta_P = \Sigma_Q - \theta_Q$$
$$\Sigma_M + \theta_M = \Sigma_P + \theta_P$$
$$\Sigma_N + \theta_N = \Sigma_Q + \theta_Q$$

whence,

$$\theta_N - \theta_M = \theta_Q - \theta_P \tag{31}$$

This relationship is known as Hencky's first theorem. In the case where ϕ is independent of p (for the Tresca or Coulomb criterion), the theorem leads to a simple geometrical property: the variation of the tangent angle along an α characteristic line between two given β characteristic lines is independent of the considered α characteristic line (and similarly if the roles of α and β are reversed). A network having this property is called a *Hencky network*.

10.3 Consequences of Hencky's theorem for some simple fields

Semi-homogeneous field: In the conditions for which Hencky's theorem is applicable the stress field is said to be semi-homogeneous in a domain if α (resp. β) in equations (27) is constant in that domain.

By reference to Figure IV.5 again, it may be seen that θ and Σ are constant on each α (resp. β) line in the domain; also, the α (resp. β) characteristics are rectilinear in that domain.

Homogeneous field: In the conditions for which Hencky's theorem is applicable, the stress-field is said to be homogeneous in a domain if both α and β are constant. Then Σ and θ are constant in the domain, and both families of characteristic lines are rectilinear.

10.4 Some properties of semi-homogeneous fields

It is of interest to deal in more detail with semi-homogeneous fields for the Tresca or Coulomb material, as they appear to be very useful for obtaining solutions.

The α (resp. β) characteristic lines of such a field are segments of a family of straight lines depending on one parameter.[1] [Reciprocally, it can be proved that any family of straight lines depending on one parameter may be taken as a family of α (or β) lines.] These straight lines have an envelope, and it can be shown that points where characteristics come into contact with the envelope cannot lie within the body.[2]

An important case is that where the envelope converges to a single point, forming what is known as Prandtl's fan [47]. This point appears as a degenerated β (resp. α) characteristic. If the Coulomb criterion is used, then Prandtl's centred

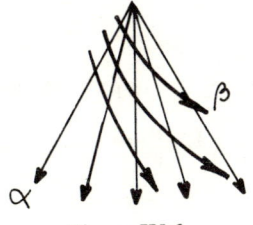

Figure IV.6

---

[1] One may take, for instance, the curvilinear abscissa along a given β (resp. α) line.
[2] This result is from a particular case of Bonneau's theorem (3)], given in Appendix B of Chapter V.

fan is constituted by logarithmic spirals (circles in the case of the Tresca criterion) and radii (see Figure IV.6).

In the general case for the Tresca criterion, where the characteristic network is orthogonal, the β (resp. α) characteristics are the involutes of the envelope (E) (Figure IV.7). The abscissa on (E) of the point of contact T of each α line with (E) is denoted by s and the distance TM from this point of contact to the corresponding point on a β line by r (Figure IV.7). It is known that $\rho = s + r$ is a constant along each β involute of (E).

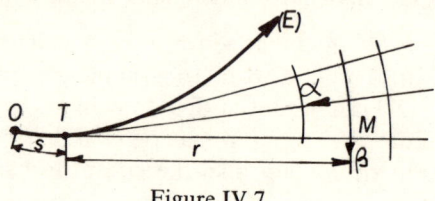

Figure IV.7

Hence, all the rectilinear segments comprised between two slip lines of the other family have the same length.

11. Matching of Solutions

As the characteristic lines are real, it is possible[1] to match two solutions *along such a line* which correspond to the same values of p and θ, satisfying (24) and (25), but differing in the values of their derivatives. Thus, a continuous solution is obtained, with discontinuous derivatives normal to the characteristic.

If Hencky's theorem is applicable, it is easy to see that only a semi-homogeneous field can be matched with a homogeneous field. If the matching is carried out along an α (resp. β) line it is the α (resp. β) parameter that is constant in the semi-homogeneous field: for the Tresca or Coulomb criterion, it is therefore the α lines (resp. β) which are straight in the semi-homogeneous field.

The results obtained so far refer to the stress field in the plastic zones. Although the stress problem has not yet been fully explored, an investigation of the problem of the velocities will now be commenced. The problem of discontinuous stress fields in the plastic zone will be dealt with at the end of Part B (Section 16). An example will be given in Part C, and in Part D some peculiarities of problems involving a Coulomb material will be mentioned.

B. THE VELOCITY PROBLEM

12. Flow Rule

12.1

The participation of the flow rule has been dealt with in Sections 2 and 3. For materials with a criterion of the intrinsic curve type it was found to be

---

[1] The indeterminacy of the solution of Cauchy's problem on a slip line arc with characteristic data arises from the indeterminacy in the calculation of the normal derivatives.

unnecessary for the flow rule to be that of the standard material in order to have the stress problem in the plastic zones formulated in the manner indicated. Thus, it was possible to deal with this problem without fully specifying the flow rule, provided that certain conditions were satisfied. On the contrary it is obvious that the flow rule must be thoroughly defined for the study of the velocity problem.

As a simplification, only the case of a standard Tresca material is to be investigated. It is known that the behaviour pattern in such a case is acceptable for metal and undrained clays ($\phi = 0$).

The case of a standard material with any type of intrinsic curve is dealt with in Appendix A. Although the hypothesis of the principle of maximum plastic work is not physically realistic in the case of $\phi \neq 0$, it is useful to deal with the problem under this hypothesis:

(1). On the one hand, the solution for the standard material can supply information with respect to the behaviour of the real material (cf. in Chapter V, Radenkovic's theorem, [48]);

(2). On the other hand, as indicated in the Appendix, the study of the velocity problem for a standard material with any type of intrinsic curve supplies all the information required to determine solutions for non-standard materials which, in plane problems, have both a yield criterion and a plastic potential of the intrinsic curve type (see, for example, [44]).

12.2 The standard material

In the case of a standard material, the flow rule is

$$v_{ij} = \lambda \frac{\partial F}{\partial \sigma_{ij}}, \lambda \geqslant 0 \tag{32}$$

where

$F(\sigma_x, \sigma_y, T_{xy}, \sigma_z) = 0$ is the yield criterion;

$f(\sigma_x, \sigma_y, \tau_{xy}) = 0$ is the projection of this criterion onto the plane (x, y)

so that

$$f(\sigma_x, \sigma_y, \tau_{xy}) = F[\sigma_x, \sigma_y, \tau_{xy}, \sigma_z(\sigma_x, \sigma_y, \tau_{xy})]$$

where the function $\sigma_z(\sigma_x, \sigma_y, \tau_{xy})$ is obtained by resolving

$$\frac{\partial F}{\partial \sigma_{xz}} = 0$$

Therefore

$$\frac{\partial f}{\partial \sigma_{ij}} = \frac{\partial F}{\partial \sigma_{ij}} + \frac{\partial F}{\partial \sigma_z}\frac{\partial \sigma_z}{\partial \sigma_{ij}} = \frac{\partial F}{\partial \sigma_{ij}} \text{ for } i,j = 1, 2 \tag{33}$$

whence

$$v_{ij} = \lambda \frac{\partial f}{\partial \sigma_{ij}} \text{ for } \lambda \geq 0, i, j = 1, 2 \tag{34}$$

Thus f, the 'criterion in the plane (x, y)', is also the two dimensional plastic potential.

For the case where F depends only on the deviator,

$$f(\sigma_x, \sigma_y, \tau_{xy}) = \frac{(\sigma_x - \sigma_y)^2}{4} + \frac{\tau_{xy}^2}{2} + \frac{\tau_{yx}^2}{2} - k^2 \tag{35}$$

where

$$\left.\begin{aligned} v_{xx} &= \lambda(\sigma_x - \sigma_y)/2 \\ v_{yy} &= \lambda(\sigma_y - \sigma_x)/2 \\ v_{xy} &= \lambda\tau_{xy} \\ \lambda &\geq 0 \end{aligned}\right\} \tag{36}$$

13. Characteristics—Relations for the Velocities Along the Characteristics

With the stress problem being solved as indicated in Part A, i.e. the stresses being determined in the plastic zones, the corresponding velocities are now required.

The zones in which σ was not determined are assumed *a priori* to be undeformed zones, in which the yield state is not necessarily reached. The motion is a rigid-body motion, and the velocity field is determined from the boundary conditions. Interest will now concentrate on the velocity distribution within the plastic zones.

As only a Tresca material is being investigated, the α and β characteristic lines are orthogonal.

At a point M of the plastic zone, let Mx and My be axes tangential to $M\alpha$ and $M\beta$ (Figure IV. 8). Then, at M, $\sigma_x = -p = \sigma_y, \tau_{xy} = k$

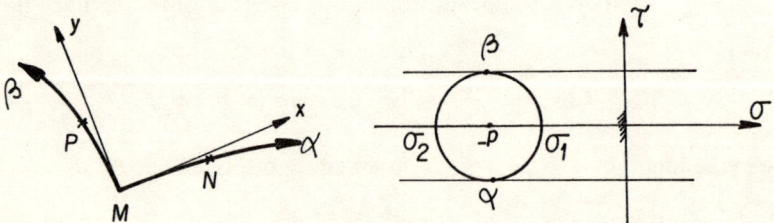

Figure IV.8

The flow rule (36) yields

$$\frac{(\partial u_x/\partial x)}{0} = \frac{(\partial u_y/\partial y)}{0} = \frac{1}{2}\frac{(\partial u_x/\partial y) + (\partial u_y/\partial x)}{k} = \lambda \geq 0 \tag{37}$$

From this we derive the following result. *At a point M the extension rate following the directions of Mα and Mβ is zero and the shear strain rate must be positive.*

With the introduction of v_α and v_β, the components of the velocity following directions α and β, equation (37) can be expressed, by equating the projections on Mx of the velocities at M and N (also at M and P on My), as

$$dv_\alpha - v_\beta \, d\theta = 0 \text{ along } \alpha \text{ lines}$$
$$dv_\beta + v_\alpha \, d\theta = 0 \text{ along } \beta \text{ lines} \qquad (38)$$

and the necessary condition ($\lambda \geq 0$) as

$$\frac{\partial v_\alpha}{\partial s_\beta} - v_\beta \frac{\partial \theta}{\partial s_\beta} + \frac{\partial v_\beta}{\partial s_\alpha} + v_\alpha \frac{\partial \theta}{\partial s_\alpha} \geq 0 \qquad (39)$$

From a mathematical standpoint, the problem of the velocities in the plastic zones where the stress-field is known, as defined by (36), is hyperbolic and linear. The α and β characteristics for the stresses are shown by (38) to be the characteristics for the velocities.[1] These equations, referred to as Geiringer's equations (17, 18), are the relations along the characteristics. Geiringer's equations state that the deformation in the plane (x, y) occurs without volume change, and has as (orthogonal) directions of zero extension the directions of the α and β stress characteristics.

14. Examples

Normally the determination of the velocity field in the plastic zone is carried out using the method of characteristics. It is possible in some cases to determine the explicit form of the solution. An example is given by the velocity fields associated with the simple stress fields met in Section 10 [36, 38].

14.1 Homogeneous stress-field

The characteristics constitute a network of orthogonal lines that can be taken as lines of cartesian coordinates: $v_\alpha = u_x$, $v_\beta = v_y$ (see Figure IV.9). Equation (38) shows that

$$\frac{\partial u_x}{\partial x} = \frac{\partial u_y}{\partial y} = 0$$

Hence, the form of the general solution for the velocity field in a homogeneous stress field is

$$u_x = f(y), \; u_y = g(x) \qquad (40)$$

and expression (39) yields

$$f'(y) + g'(x) \geq 0 \qquad (41)$$

---

[1] The characteristics of such a problem of the first order are the curves along which a differential relation exists for the solution.

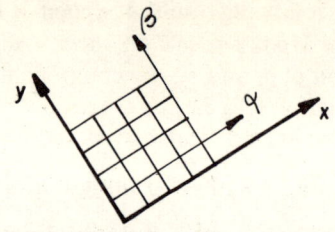

Figure IV.9

14.2 Semi-homogeneous stress-field

The β characteristic lines are assumed to be rectilinear segments whose envelope is (E) (Figure IV.10).

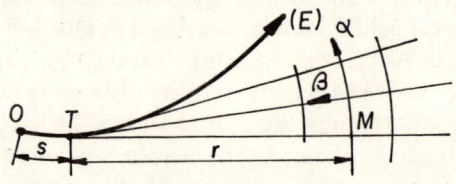

Figure IV.10

The Geiringer equations become

$$dv_\alpha - v_\beta \, d\theta = 0 \text{ along } \alpha \text{ lines}$$
$$d v_\beta = 0 \text{ along } \beta \text{ lines} \tag{42}$$

The α and β characteristic lines can be taken as a system of curvilinear orthogonal coordinates, the corresponding parameters being θ, constant on the β lines and variable on the α lines, and p, constant on the α lines and variable on the β lines (cf. Section 10.4).

Equation (42) implies that v_β is a function of θ only; for instance,

$$v_\beta = f'(\theta) \tag{43}$$

and since

$$\frac{\partial v_\alpha}{\partial \theta} = v_\beta$$

it follows that

$$v_\alpha = f(\theta) + g(p) \tag{44}$$

Equations (43) and (44) give the general form of the solution for the velocities in a semi-homogeneous stress field.

The necessary condition (39) is, therefore, (since $dS_\beta = -dp$ and $ds_\alpha = r \, d\theta$) written as

$$-g'(p) + \frac{f''(\theta)}{r} + [f(\theta) + g(p)]\frac{1}{r} \geq 0 \tag{45}$$

15. Discontinuity of the Velocity

The problem as dealt with so far has concerned continuous velocity fields in a plastic zone for which the stress field is known and is continuous. Next comes the problem where the velocity field is discontinuous, although the stress field is continuous (the determination of the weak solutions).[1]

As the stress-field is known, the characteristic lines network is fixed. This corresponds to the fact that, with the stresses being determined, the problem for the velocities is linear.

It is known (see Courant and Hilbert [11] II pages 486 and following) that for any linear hyperbolic problem the discontinuity lines of the weak solutions are necessarily characteristic lines. A jump condition and an equation for the propagation of the discontinuity along the characteristic line are obtained.

The main results may be summarized as follows.

(1). All the velocity jump lines are necessarily characteristic lines.

(2). As the velocity flux through the discontinuity line must be conserved, because of the incompressibility of the material, the *velocity jump is necessarily tangential*. This is the very jump condition foreseen above.

(3). If, for instance, the line of discontinuity of **u** is an α line, then in the close neighbourhood of this line Geiringer's equation can be applied: with $dv_\alpha - v_\beta\, d\theta = 0$ following the α line and v_β being continuous across the α line, the propagation equation results; i.e.[2]

$$d[v_\alpha] = 0.$$

The velocity jump is constant along the discontinuity line.

(4). Finally, the necessary conditions corresponding to $\lambda \geqslant 0$ mean that, if the α (resp. β) line is crossed in the direction of the β (resp. α) line, *the discontinuity of v_α (resp. v_β) must be positive*.

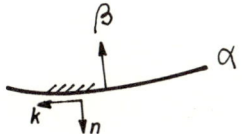

Figure IV.11

It will be noticed that the characteristic lines which are the maximum shear stress lines appear as the possible lines of sliding. From this comes the fact that α and β lines are usually termed *slip-lines*. This is in agreement with some experimental results in Soil Mechanics [2, 22] (Figure IV.11).

---

[1] The need for taking into account the weak solutions of the mathematical problem appears in practically all examples. This is clearly suggested by experiments as one often observes the localization of the deformation in very thin zones which may be considered as sliding along a surface.

[2] Square brackets indicate the discontinuity.

16. Discontinuity of the Stress-field

As stated in Section 11, this section deals with the weak solutions for stresses that are sometimes necessary in a particular type of problem. (See, for example, [63].)

The stress problem is quasi-linear and takes the form of a conservation law. The mathematical study is given in [11] II pages 488 to 490, the results of which are as follows.

(1). The discontinuity lines are *not* characteristics.
(2). Weak solutions can be obtained even if the data are continuous.
(3). Jump conditions are obtained, which correspond to the continuity of the stress vector applied to the discontinuity surface. These are not sufficient to determine a unique discontinuous solution, and in practice it is necessary to have some indication of the shape of the discontinuity line.

With regard to the velocities, a velocity discontinuity cannot exist along a stress discontinuity line. This line must be considered as the boundary of an infinitely thin rigid zone. (Cf. Reference [26] *and Appendix* A, Section 7.2, of the present chapter[1].)

C. STUDY OF AN EXAMPLE

17. The Problem

An example is now given to demonstrate the concepts introduced in the preceding paragraphs. The problem considered, as in (38), is the passive pressure of a weightless plastic soil on a smooth rigid plate. The soil is assumed to obey the Tresca criterion.

The boundary conditions are as follows (Figure IV.12). At infinity, the velocities are zero.

On OY, $\quad \sigma = -q, \tau = 0$

On OX, $\quad \begin{cases} s > 0: \tau = 0, u_n = \Omega(a + s) \\ s < 0: \tau = 0, \sigma = 0 \end{cases}$

$IM = a + s$

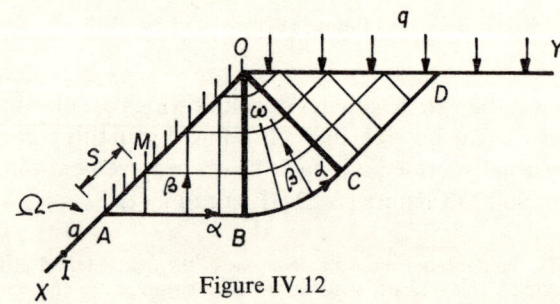

Figure IV.12

---

[1] This is what makes the weak solutions for stresses significant with respect to the physical problem.

18. Construction of the Solution

It is intended to construct a solution corresponding to an upward flow of soil, i.e. the way the plastic zone between AO and OY would move.

By starting along OY and applying the method of Section 8 it is seen that in the angle $(OC, OY) = \pi/4$ there exists a homogeneous field, as $\tau = 0$ and σ = constant on OY. Two such equilibrium situations are possible, because OC can either be an α or β characteristic line (two possible Mohr circles). Therefore the velocities must intervene (and, therefore, the flow rule): as the motion of the material must proceed from OA towards the surface OY, the principal stress σ must be the greatest traction and therefore OC is a β line.[1]

Only a semi-homogeneous field can be connected to a homogeneous field, where the β characteristics will be straight lines. Any β line of this semi-homogeneous field that cuts AO must make an angle of $\pi/4$ with $AO (\tau = 0$ on $AO)$. These lines will, therefore, be parallel straight lines and we have under AO another homogeneous field, OAB.

There remains the angle BOC, which connects both homogeneous fields. If this is assumed to be continuous (in order to have a continuous solution), then the area may be sub-sectioned using a Prandtl's fan centred[2] in O with an aperture ω.

The stress on OA is normal and uniform and is simple to calculate. Its value is determined using the relation along α characteristic lines from OY to OX. On OY,

$$p = -\sigma_1 + k = q + k$$

whence, on OA

$$p = (q + k) + 2k\omega$$

and the pressure on OA is given by

$$-\sigma_2 = p + k = q + 2k(1 + \omega) \tag{46}$$

19. Calculation of the Velocities

Under $ABCD$, the wedge considered as rigid must be immobile in accordance with the conditions at infinity. Geiringer's equations (38) show that v_β is constant along each β line. Also v_β is continuous when crossing α lines, in particular the $ABCD$ line. As $v_\beta = 0$ under $ABCD$, $v_\beta = 0$ on $ABCD$ also, in the plastic zone. Therefore, $v_\beta = 0$ in the whole of the plastic zone.

The calculation of v_α commences along OA. In OAB v_α = constant along

---

[1] This reasoning is not perfectly rigorous, and uses some intuition; but it is not indispensable. The problem of defining the α and β characteristic lines can be left aside and the two possible equilibrium situations that supply two distinct values for the pressure on OA can be studied. The question is settled when the velocity problem is considered. Equation (39) must be satisfied at any point, and this appears to be possible for only one selection of α and β (at the most, since it may happen that neither selection makes it possible to satisfy (39) everywhere).

[2] O is not within the plastic range, but is on the boundary, in agreement with Section 10.

each α line (from equations (38)) and on OA, $u_n = \Omega(a + s)$. Since $v_\beta = 0$, there is soil sliding underneath OA (which is admissible as OA is smooth) from A towards O, and

$$v_\alpha = \Omega(a + s)\sqrt{2} \qquad (47)$$

With the notation of Section 14.1 (Figure IV.13),

$$v_\alpha = \Omega(a\sqrt{2} + 2y) = f(y)$$

from (40) and it is verified that $f'(y) \geq 0$.

In OBC, v_α = constant along each α line according to equation (38), and the continuity of v_α, when crossing OB, determines its value.

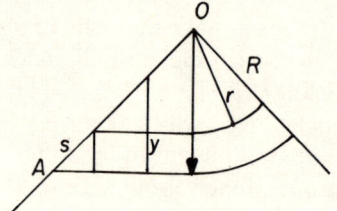

Figure IV.13

Hence $v_\alpha = \Omega(a\sqrt{2} + 2(R - r)) = g(p) = g(r)$ with the notations of Section 14.2. A check is made that $g(p) - rg'(p) \geq 0$.

In OCD, the field is identical to that in OAB.

Finally, there is a discontinuity of the velocity along $ABCD$. $[v_\alpha] = \Omega a\sqrt{2}$ is positive when $ABCD$ is crossed in the direction of the β lines.

20. A Particular Case

In Figure IV.14, the particular case is shown where $\omega = \pi/2$, $q = 0$ and I is at infinity, i.e. the plate is horizontal and is given a vertical translation. The plastic flow can occur on the right of A as well as on the left of A' since I is at infinity.

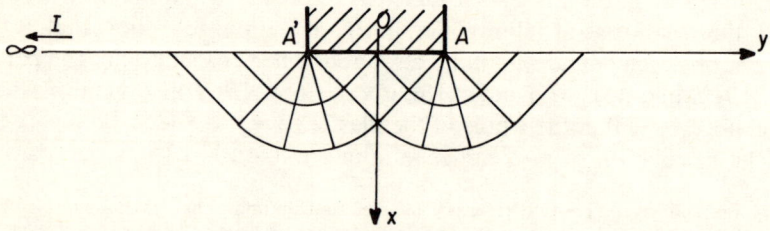

Figure IV.14

The pressure is given by

$$-\sigma_2 = (\pi + 2)k$$

This solution is valid regardless of the friction conditions beneath the plate. The symmetry of the problem requires that the equilibrium under the plate is homogeneous. (If there is friction, the velocity distribution is symmetric, but if there is no friction, the velocity distribution depends on an arbitrary non-decreasing function of y, namely the sliding velocity of the soil underneath the plate.)

21. Remarks on the Solution

For the general case of Section 19, the solution is valid only if q is not too great. Intuitively, it is seen that if q becomes too great the material tends to collapse on the left-hand side of A towards AX.

The above solution is not complete as stresses in the rigid zone under $ABCD$ have not been considered. To fully solve the problem, it is necessary to determine the stress-field that satisfies the equilibrium equations, and the boundary conditions for the stresses, and does not violate the yield criterion. It is sufficient to prove that it is possible to find such a field (which, further on, will be called an admissible stress field).[1]

The theory of limit analysis, which will be presented in the following chapter, will make it possible to give a clear significance to the results obtained.

22. The Case of a Coulomb Material

The same problem is now studied for the case of a weightless Coulomb material. In accordance with Sections 2, and 3, only the stress-field is considered, as this does not require a fully specified flow rule. (Appendix A contains all the information necessary for studying the stresses and the velocities. It should be noted that in the classical works dealing with a Coulomb material (eg. [55] and [58]), the velocity field is usually not taken into account.

The reasoning required for the construction of the solution will not be repeated in detail. The drawing for the problem is modified as indicated in Figure IV.15. The characteristics are not orthogonal in this case.

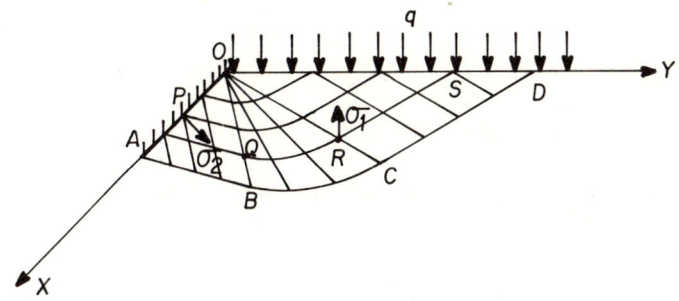

Figure IV.15

---

[1] It has been proved that this can actually be done if q is not too great [51].

As for a Tresca material, the stress applied on OA is a principal stress (as on OY). It is necessary to know whether this stress is major or minor (i.e. σ_1 or σ_2). In Section 17, this was made known by the participation of the velocity field. In this problem the movement of the plate is into the soil, and hence the stress on OA is necessarily σ_2.

Mandel [38] proposed that in the case of a Coulomb material when the flow rule is left unspecified, the selection of the solution should be based on the solution for the same problem in the case of a Tresca material with associated flow rule. In classical Soil Mechanics textbooks either the question is never posed or it is solved by 'physical' considerations via the concepts of 'maximum' or 'minimum equilibrium' of active and passive pressure. As explained later in Chapter V the choice can be made only by using the theorems *of limit analysis for standard and non-standard materials*. The real significance of the two above interpretations is that the Coulomb material is assumed to be a standard one [53].

In the present case, we derive that in OCD the field is homogeneous and $\sigma = -q = \sigma_1$, whence $(OC, OY) = \pi/4 - \phi/2$. In OAB the field is homogeneous and the pressure n under the plate is the greatest pressure, i.e. $n = -\sigma_2$ and also

$$(OA, OB) = \pi/4 + \phi/2$$

This zone is in a state of *passive limit equilibrium*.

The field in OBC is semi-homogeneous (logarithmic spirals). The stress on OA is normal and uniform. It is calculated using the relations along the characteristics.

At S on OY, $-p = \sigma_1 - R$. Hence $p = q + (H + p) \sin \phi$ and, from equation (29),

$$\Sigma_S = -\frac{\cot \phi}{2} \log (H + p) = -\frac{\cot \phi}{2} \log \frac{H + q}{1 - \sin \phi}$$

At P on OA $-p = \sigma_2 + R$. Hence $p = n - (H + p) \sin \phi$ and

$$\Sigma_p = -\frac{\cot \phi}{2} \log \frac{H + n}{1 + \sin \phi}$$

$$\Sigma_p = \Sigma_Q; \theta_p = \theta_Q$$

Applying equation (27) ($\Sigma - \theta$ = constant) along the α lines, gives

$$(n + H) = (q + H) \frac{1 + \sin \phi}{1 - \sin \phi} \varepsilon^{2\omega \tan \phi} \tag{48}$$

For the case where $\omega = \pi/2$ (the problem of the bearing capacity of a footing on a weightless soil) the solution is given by

$$P_{\text{ult}}, = -H + (q + H) \frac{1 + \sin \phi}{1 - \sin \phi} \varepsilon^{\pi \tan \phi} \tag{49}$$

which contains the terms for surface load and cohesion.

The previous arguments in Section 20 are still valid here.

In order to have a complete solution for the stresses, an admissible stress-field under *ABCD* is required.

The velocity field is not constructed, and as will be shown in Chapter V it is impossible to interpret the results obtained through this type of solution. As indicated previously, it is possible to give an interpretation by taking as a basis the solution of a similar problem for a Tresca material with associated flow rule for which the velocity field is known. In actual fact this procedure is equivalent to assuming the Coulomb material under consideration to be standard. The theory of limit loads for non-standard material [48] will be explained in Appendix A of Chapter V and will then make it possible to draw some conclusions for this type of material.

D. PARTICULAR USES IN THE STUDY OF A COULOMB MATERIAL

23. Method of Superposition—Theorem of the Corresponding States

23.1 Method of superposition

The solution process developed for the general case (Sections 7 and 8) makes it possible to study directly by a single calculation the problem of a cohesive Coulomb material with self-weight and with or without surface load.

Usually, the calculation of bearing capacities or coefficients of active and passive pressure is executed in parts, applying the so-called method of superposition. For instance, if the bearing capacity is required for a footing on a cohesionless soil ($c = 0$) having $\phi \neq 0$ with self-weight and with a surface load, then these loads are studied separately.

(a) *The problem of the bearing capacity of a material with self-weight and without a surface load.*

A stress-field σ^1 is found which satisfies the equations of equilibrium with the body forces and the boundary conditions, with imposed stresses being zero at the surface. Also,

$$f(\sigma^1) = 0 \quad \text{in the plastic zone}$$

$$f(\sigma^1) \leqslant 0 \quad \text{outside the plastic zone.}[1]$$

(b) *The problem of the bearing capacity of a weightless material with a surface load.*

A stress-field σ^2 is found which satisfies the equations of equilibrium without body forces and the boundary conditions, with stress on the surface being equal to the surface load. Also

$$f(\sigma^2) = 0 \quad \text{in the plastic zone (which can be different from that of } \sigma^1\text{);}$$

$$f(\sigma^2) \leqslant 0 \quad \text{outside the plastic zone.}[1]$$

---

[1] As already stated, the stresses are not studied outside the plastic zones. The heuristic hypothesis is made that the stress-field can be extended in each rigid zone while respecting the equilibrium equations and the boundary conditions, without violating the yield criterion. This omission is important, and will be studied later in Chapter V.

Then, the stress field $\boldsymbol{\sigma} = \boldsymbol{\sigma}^1 + \boldsymbol{\sigma}^2$ satisfies the equilibrium equations with the body forces, the boundary conditions with the surface load, and, as a consequence of the form of f,

$$f(\boldsymbol{\sigma}) = f(\boldsymbol{\sigma}^1 + \boldsymbol{\sigma}^2) \leqslant 0, \quad \text{everywhere}$$

(The elastic range of the material, $f(\boldsymbol{\sigma}) < 0$ is a convex cone with a summit O). At a given point, $f(\boldsymbol{\sigma}^1 + \boldsymbol{\sigma}^2) = 0$ only if $f(\boldsymbol{\sigma}^1) = f(\boldsymbol{\sigma}^1) = 0$; and, if both tensors $\boldsymbol{\sigma}^1$ and $\boldsymbol{\sigma}^2$ have the same ordered principal directions and therefore the same marginal plane facets, $|\tau| = h(\boldsymbol{\sigma})$. Here $\boldsymbol{\sigma}$ is known as a field in 'safe' or limit equilibrium.

Thus, the bearing capacity obtained by superposition is an approximation to the real bearing capacity and borders on the conservative side.

The method of superposition is also valid for a Coulomb material with cohesion. In this case the criterion f', acting for $\boldsymbol{\sigma}^1$, is the criterion of the corresponding cohesionless material, and for $\boldsymbol{\sigma}_2$ f is the criterion of the material with cohesion. Then,

$$f'(\boldsymbol{\sigma}^1) \leqslant 0 \quad \text{and} \quad f(\boldsymbol{\sigma}^2) \leqslant 0 \Rightarrow f(\boldsymbol{\sigma}^1 + \boldsymbol{\sigma}^2) = f(\boldsymbol{\sigma}) \leqslant 0$$

the equality $f(\boldsymbol{\sigma}) = 0$ being obtained only if $f(\boldsymbol{\sigma}^1) = f(\boldsymbol{\sigma}^2) = 0$, $\boldsymbol{\sigma}^1$ and $\boldsymbol{\sigma}^2$ having the same ordered principal directions.

23.2 Theorem of corresponding states

If the material has cohesion, the theorem of corresponding states is applied, making it possible to reduce the problem to that of a cohesionless material with a confining pressure H at the boundary.

Introducing a tensor $\boldsymbol{\sigma}'$ derived from $\boldsymbol{\sigma}$ according to

$$\boldsymbol{\sigma}' = \boldsymbol{\sigma} - H\mathbf{1}$$

reduces the problem to one in which the same equilibrium equations, with the body forces and the boundary conditions modified by adding a normal pressure equal to H, must be satisfied. The Coulomb yield criterion applies without cohesion.

For the calculation of bearing capacities, this amounts to the application of a fictitious surface load which is taken into account just as any surcharge. The confining pressure H is then subtracted from the calculated pressure.

The utilization of the theorem of corresponding states is evidently not essential. It is possible, as demonstrated in Section 22, to carry out calculations with both surface load and cohesion for a weightless material. The connection between the two terms is self-evident in the formulae. Nevertheless, this theorem is very useful to perform reductions in the formulae, *a priori*, before making use of dimensional analysis considerations, for example.

23.3 Consequences

The application of the method of superposition leads, in the case of calculations of bearing capacity, to the classical relationship due to Terzaghi:

$$P_{\text{ult}} = \frac{F}{B} = qN_q + CN_c + \tfrac{1}{2}\gamma B N_\gamma \qquad (50)$$

where N_γ, N_q, and N_c are, in the case of a homogeneous material, functions of ϕ only. With charts of N_γ, N_q and N_c as functions of ϕ available, it is possible to calculate P_{ult} for any values of q, C, γ and B.

From the practical viewpoint, this relationship has advantages. In fact, in the global calculation, the different parameters are mixed, giving through considerations of dimensional analysis, a relation for P_{ult} of the form,

$$R(P_{\text{ult}}/\gamma B,\ q/\gamma B,\ C/\gamma B,\ \phi) = 0$$

By applying the theorem of corresponding states

$$P_{\text{ult}} = q + (C + q \tan \phi) \cdot N_1\left(\frac{\gamma B}{C + q \tan \phi},\ \phi\right) \qquad (51)$$

or, equivalently,

$$P_{\text{ult}} = -C \cot \phi + (q + C \cot \phi) \cdot N_2\left(\frac{\gamma B}{q + C \cot \phi},\ \phi\right)$$

For the application of these formulae it would be necessary to have at one's disposal the curves representing N_1 or N_2 as functions of their arguments, for each value of ϕ. However, computations by the global method carried out under the direction of the author [16, 50, 53] have shown that, through a convenient reduction of the coordinates, it is possible to use equation (51) by referring to a single curve for $4° \leq \phi \leq 40°$, apart from the charts giving N_γ and N_q. In fact, for $4° \leq \phi \leq 40°$ the curves representing

$$\frac{P_{\text{ult}} + C \cot \phi}{N_q(q + C \cot \phi)}$$

as functions of

$$\frac{N_\gamma}{N_q} \cdot \frac{\gamma B}{2(q + C \cot \phi)}$$

constituted a very narrow band. Thus, the utilization of the global calculation for the bearing capacity of a footing no longer appears impossible from the practical viewpoint.

23.4 Remarks

The proof of the theorem of corresponding states, in order to be rigorous, ought also to consider the velocity fields. For this purpose it would be necessary for the constitutive law for a Coulomb soil to be specified, which is not normal practice in Soil Mechanics. Likewise, attention must also be given to the friction conditions (roughness) between the soil and, for example, a footing.

Generally, the stated results are true if the flow rules of the materials, with or without cohesion, are identical for corresponding states. (This is actually the case for all the flow rules currently proposed for soils.)

On the other hand, as regards the proof of the superposition method, the latter is supported by the concept that, since the $\sigma^1 + \sigma^2$ field is 'safe', the corresponding bearing force (for the example of a footing problem) is necessarily an under-estimation (in the sense of safety) of the real bearing capacity. It will be seen in Chapter V that this apparently intuitive idea is but the static theorem of the theory of limit analysis, the validity of which has been proved only in the case of a standard material. Stated otherwise, the proof of the method of superposition implicitly assumes that the considered Coulomb material is standard.

24. An Example Study of a Cohesionless Soil with Self-weight

In order to demonstrate some peculiarities of the problem of a cohesionless Coulomb material with self-weight, the example chosen is the passive pressure caused by a smooth plate acting on a wedge (Figure IV.16). The ground surface is stress free and inclined at an angle β to the horizontal. The wall has a batter equal to λ.

This problem must be associated with that of Section 22 in the application of the method of superposition.[1]

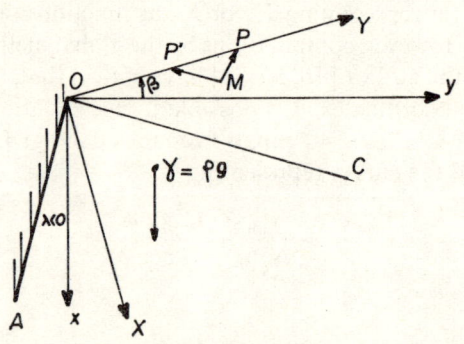

Figure IV.16

The boundary conditions are that

$$\text{on } OY, \ \sigma = \tau = 0;$$

$$\text{on } OX, \ \tau = 0, \sigma = ?$$

Strictly speaking, conditions on the velocities cannot be usefully imposed as a Coulomb material is assumed. However, as the equilibrium is that of a passive

---

[1] In the case of Section 22, the inclination of the stress-free surface ($\beta \neq 0$) would involve no complication with respect to the proposed solution.

pressure the motion of the smooth wall OA is assumed to be towards the right-hand side. A solution analogous to that of Sections 18 and 22 is sought for, starting from OY.

A peculiarity appears at once: $\sigma = \tau = 0$, whence $p = 0$, and, therefore, the Mohr circle of limit equilibrium has a zero radius R along OY. Thus, on OY only the value of p is known, as θ is not given. In fact, it has been proved [15] that the value of θ is fixed.

The stress-free surface is a line of singular points at which the determinant involved in the solution of Cauchy's problem is zero. The problem has a solution only if a condition which determines the value of θ is fulfilled. This condition can be explained as follows.

Assuming that θ is known on OY, and two points P and P' infinitely close to one another are considered, an attempt is made to apply this method of characteristics (Figure IV.16). Equations (24) and (25) are discretized, taking the value of R at the surface to be zero, and this leads to two linear equations for only one unknown, the value of p at M. Such equations must be compatible.

More rigorously, along OY d$p = 0$, or

$$-\frac{\partial p}{\partial x} \sin \beta + \frac{\partial p}{\partial y} \cos \beta = 0$$

From (17) and (18),

$$\frac{\partial p}{\partial x} = \gamma \frac{1 + \sin \phi \cos 2\theta}{\cos^2 \phi}, \quad \frac{\partial p}{\partial y} = \gamma \frac{\sin \phi \sin 2\theta}{\cos^2 \phi}$$

whence,

$$\sin (2\theta - \beta) = \frac{\sin \beta}{\sin \phi}$$

The possible values of θ are either

$$(Ox, \sigma_1) = \theta = \pi + \frac{\beta}{2} + \frac{1}{2} \text{Arc sin} \frac{\sin \beta}{\sin \phi} \tag{52}$$

or

$$\theta = \frac{\pi}{2} + \frac{\beta}{2} - \frac{1}{2} \text{Arc sin} \frac{\sin \beta}{\sin \phi} \tag{53}$$

The choice between these two possible values corresponds to that which has already been operative in the cases of a Tresca material and a weightless Coulomb material. Thus, the indeterminacy is removed as indicated in Section 22. The rule is that for a passive pressure the solution is that corresponding to the greatest force on the wall, i.e. expression (52), in which σ_1 is closer to the normal on the stress-free surface than in relation (53). The latter expression corresponds, therefore, to the case of active pressure.

The field is now known in the domain of determinacy of OY. There appears to be an indeterminacy as the value of the normal derivative of θ along OY can be arbitrarily chosen, since in expressions (17) and (18) $\partial\theta/\partial x$ and $\partial\theta/\partial y$ have zero co-efficients.

In fact this value is fixed, and is necessarily zero. To prove this it is convenient to write (17) and (18) using axes OX and OY. Then $\partial\theta/\partial y$, $\partial p/\partial y$ and the derivatives of higher order with respect to Y are zero on OY. Equations (17) and (18) must be satisfied within the whole zone in limit equilibrium. Therefore, from the equations derived from (17) and (18) by differentiation,

$$\frac{\partial\theta}{\partial X} = 0 \tag{54}$$

In this field, θ is constant, the characteristics are straight lines and the stresses depend only on X: the zone within the angle (OY, OC) is in *Rankine equilibrium*.

The solution for the angle (OC, OA) must now be determined. This can be done using the known values of p and θ on the OC characteristic and a known relationship for θ on the non-characteristic line OA:

$$\theta = \pi + \lambda \text{ (for a smooth plate)}$$

It should be observed that O is necessarily a point of discontinuity of the solution, as θ has two distinct values at that point, given by expression (52) and by $\theta = \pi + \lambda$. For Tresca and Coulomb weightless materials this singularity of the boundary conditions has been satisfied by the introduction of Prandtl's fan, centred at a point O in the solution. Here a property of invariance by homothetia with respect to point O will be used. A solution in the form,

$$p = \gamma r S(\omega), \quad \theta = \theta(\omega) \tag{55}$$

where r and ω denote polar coordinates with respect to Ox, will be shown to be suitable.

For the cohesionless Coulomb material in the plastic zone,

$$R = p \sin \phi$$

whence, by means of (13),

$$\left.\begin{array}{l} \sigma_r = -p(1 - \sin\phi \cos 2(\theta - \omega)) \\ \sigma_\omega = -p(1 + \sin\phi \cos 2(\theta - \omega)) \\ \tau_{\omega r} = p \sin\phi \sin 2(\theta - \omega) \end{array}\right\} \tag{56}$$

As p and θ have the form (55), the boundary conditions on OC and OA are satisfied, being homothetic with respect to O. Here the angle θ is independent of r, and the stresses on OC are proportional to r (as shown by integration of (24) on OC). The equilibrium equations in polar coordinates yield

$$\left.\begin{array}{l} \dfrac{\partial \sigma_r}{\partial r} + \dfrac{1}{r}\dfrac{\partial \tau_{r\omega}}{\partial \omega} + \dfrac{\sigma_r - \sigma_\omega}{r} + \gamma_r = 0 \\[2mm] \dfrac{\partial \tau_{r\omega}}{\partial r} + \dfrac{1}{r}\dfrac{\partial \sigma_\omega}{\partial \omega} + \dfrac{2\tau_{r\omega}}{r} + \gamma_\omega = 0 \end{array}\right\} \tag{57}$$

where

$$\gamma_r = \gamma \cos \omega$$

$$\gamma_\omega = -\gamma \sin \omega$$

It can, therefore, be seen that the form of solution (55) is admissible. The terms in r vanish, and (57) is reduced to the system (58) of two differential equations of the first order for $S(\omega)$ and $\theta(\omega)$

$$\frac{dS}{d\omega} = \frac{S \sin 2(\theta - \omega) - \sin (2\theta - \omega)}{\cos 2(\theta - \omega) + \sin \phi}$$

$$\frac{d\theta}{d\omega} = \frac{-\cos \omega - \sin \phi \cos (2\theta - \omega) + S \cos^2 \phi}{2S \sin \phi \left[\cos 2(\theta - \omega) + \sin \phi\right]} \quad (58)$$

Equation (58) is to be solved with the data of $S(\omega)$ and $\theta(\omega)$ on OC, and of $\theta(\omega)$ on OA, which gives, apparently, one condition too many. Actually, as OC is a characteristic and as the data of p and θ on OC are characteristic, equations (58) are indeterminate on OC as a starting line with these data, and the data on OA must be taken into account so as to determine the solution (by the so-called trial and error method). Thus, $S(\omega)$ and hence p, and $\theta(\omega)$, are obtained.

The solution derived is the only continuous solution of the problem. It is homothetic with respect to O and the distribution of stresses on the wall is triangular. The characteristic network presents the appearance shown in Figure IV.17. It is seen that in this case, unlike that of Section 22, there is no fan in O, only one β characteristic in O (i.e. OC), and O is also a point of discontinuity (49).

From the Soil Mechanics viewpoint, the solution of this problem gives the value of the weight factor in the problem of the passive pressure.

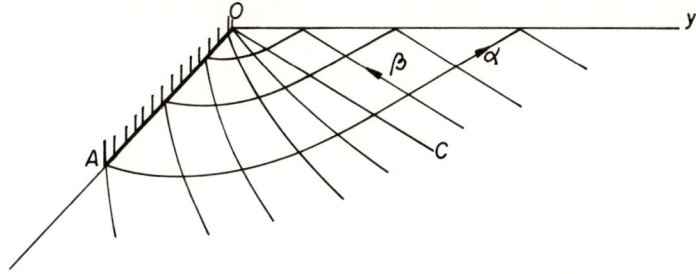

Figure IV.17

For the calculation of bearing capacity by application of the method of superposition, the N_c and N_q coefficients of (50) are given by (49). For N_y the solution must be determined for a footing on a soil with self-weight, having $\phi \neq 0$ and without cohesion. To do this, the above solution is used for the case of the aperture equal to π. The corresponding characteristics network consists, on the right-hand side, of the network of Figure IV.17 (for the aperture π), bounded by both

α and β characteristics which are bisected by the axis of the footing; and the left-hand side of the network is formed by symmetry about that axis (34). Unlike the value of N_c and N_q that of N_γ depends on the friction condition under the footing.

The global method calculations of bearing capacity carried out by Lundgren and Mortensen for $\phi = 30$ [34] showed that the solution for a material with weight, without cohesion or surface load, and in which a point of discontinuity occurs at the corner of the footing, appeared as the limit of the global solution when the cohesion and the surface load tended to zero. This was confirmed in (16) and (50) for other values of ϕ. Such matters will not be developed further as many text books deal with them in detail.

New problems are continuously being published as the appearance of efficient computer methods opens up new horizons in this field.

References

[1] B. G. Berezancew (1952) *Problème de l'Équilibre Limite d'un Milieu Pulvérulent en Symétrie Axiale,* Moscow.

[2] J. P. Boehler and A. Sawczuk (1970) Equilbre limite des sals anisotropes, *J. de Mécanique,* **9,** No. 1, pp. 5–33.

[3] M. Bonneau (1947) Equilibre limite et rupture des milieux continus, *Ann. Pts et Ch.,* Sept.–Oct. 1947, pp. 609–653, Nov–Dec. 1947, pp. 669–801.

[4] J. R. Booker (1970) A property of limiting lines for a perfectly plastic material, *Univ. Sydney, Civ. Eng. Lab., Res. Rept.,* No. R 134, March 1970.

[5] J. R. Booker (1970) An improved method for the numerical integration of hyperbolic equations, *Univ. Sydney Civ. Eng. Lab., Res. Rept.,* No. 154, Nov. 1970.

[6] J. R. Booker (1972) A method of integration for the equations of plasticity of a weightless cohesive frictional material, *Quart. Jl. Mech. Appl. Math.,* **25,** No. 1, pp. 63–82.

[7] J. R. Booker and E. H. Davis (1972) A general treatment of plastic anisotropy under conditions of plane strain, *Jl. Mech. Phys. Solids,* **20,** No. 4, pp. 239–250.

[8] A Caquot and J. Kerisel (1953) Sur le terme de surface dans le calcul des fondations en milieu pulvérulent, *Proc. 3rd. Int. Conf. Sci. Mech.,* Zürich, Vol. 1, pp. 336–337.

[9] A. Caquot and J. Kerisel (1966) *Traité de Mécanique des Sols,* 4th ed., Gauthier-Villars, Paris.

[10] I. F. Collins (1970) A slip line field analysis of the deformation at the confluence of two glacier streams, *Jl. of Glaciology,* **9,** No. 56, pp. 169–193.

[11] R. Courant and D. Hilbert (1962) *Methods of Mathematical Physics,* Vol. II, Interscience, New York.

[12] A. D. Cox, G. Eason and H. G. Hopkins (1961) Axially symmetric plastic deformations in soils, *Phil. Trans. Roy. Soc. London, A,* **1036,** 254, pp. 1–54.

[13] F. H. Davis and J. T. Christian (1971) Bearing capacity of anisotropic cohesive soil, *Jl. Soil Mech. & Found. Div., ASCE,* **97,** No. SME, pp. 753–769.

[14] Y. d'Escatha and J. Mandel (1971) Profondeur critique d'eboulement d'un souterrain, *C.R. Ac. Sc., Paris, A,* **273,** pp. 470–473.

[15] J. Ferrandon (1951) Techniques de l'Ingénieur, *Construction I,* Mécanique des Sols, p. 220.

[16] P. Florentin and Y. Gabriel (1974) Force partante d'une fondation sut sol verticalement non homogène, *Trav. fin d'études E.N.P.C.,* Laboratoire de Mécanique, Ecole Polytechnique, Paris, June 1974.

[17] H. Geiringer (1930) Beit zum Vollständigen ebenen Plastizitäts Problem, *Proc. 3rd Int. Congr. Appl. Mech.,* **2,** pp. 185–190.

[18] H. Geiringer (1937) Fondements mathématiques de la théorie des corps plastiques isotropes, *Mem. Sc. Math.*, **86**, Gauthier:Villars, Paris.
[19] J. P. Giroud, Tran vo Nhiem, and J. P. Obin (1974) *Tables pour le Calcul des Fondations*, Vol. 3, Dunod, Paris.
[20] A. P. Green (1962) Two-dimensional problems in plasticity, *Handbook of Eng. Mech.*, Ed. W. Flügge, McGraw-Hill Book, Co., New York.
[21] A. Haar and Th. Karman (1909) Zur Theorie der Spannungszustände in plastichen und sandartigen Medien, *Nachr. Ges. Wiss. Göttingen, Math. Phys. Kl.*, pp. 204–218.
[22] P. Habib (1953) Etude de l'orientation du plan de rupture et de l'angle de frottement interne de certaines argiles, *Proc. 3rd. Int. Conf. Soil Mech.*, Zürich, **1**, pp. 28–31.
[23] Bent Hansen A theory of plasticity for ideal frictionless material, Thesis.
[24] H. Hencky (1923) Über einige statisch bestimmte Fàlle des Gleichsewichts in plastischen Körpern, *Z. Angew. Math. Mech.*, 3, pp. 241.–251.
[25] R. Hill (1949) The theory of plane plastic strain for anisotropic metals, *Proc. Roy. Soc., A*, **198**, pp. 428–437.
[26] R. Hill (1950) *The Mathematical Theory of Plasticity*, Clarendon Press, Oxford.
[27] R. Hill (1967) On the vectorial superposition of Henchy–Prandt. nets, *Jl. Mech. Phys. Solids*, **15**, No. 4, pp. 255–262.
[28] Hon Yimko and L. W. Davidson (1973) Bearing capacity of footings in plane strain, *Jl. Soil Mech. and Found. Div., ASCE*, **99**, No. SM1.
[29] A. B. Jenike and R. T. Shield (1959) On the plastic flow of Coulomb solids beyond original failure. *Jl. Appl. Mech., Trans. A.C.M.E.*, **26**, No. 4, pp. 599–602.
[30] G. de Josselin de Jong (1959) *Statics and Kinematics in the failable Zone of a Granular Material*, Uitgeverij-Delft.
[31] F. Kötter (1903) Die Bestimmung des Druches an gebrümmeten Blattflächen, eine Aufgabe aus der Lehre vom Erddruck, *Berl. Akad. Bericht*, p. 229.
[32] F. Kötter (1909) Über den Druck von Sand, *Berl. Akad. Bericht*.
[33] J. Legrand (1970) *Cours de Mécanique des Sols*, E.N.P.C., Paris.
[34] H. Lundgren and K. Mortemsen (1953) Determination by the theory of plasticity of the bearing capacity of continuous footings on sand. *Proc. 3rd Int. Conf. Soil Mech.*, Zürich, **1**, pp. 409–412.
[35] J. Mandel (1942) Equilibres par tranches planes des solides à limite d'écoulement, Thesis (Louis Jean, Gap.), v.z. Travaux, Juin–Juillet–Décembre 1943.
[36] J. Mandel (1961) Problèmes de déformation plane (et de contrai plane) pour les corps parfaitement plastiques, *Sem. Plasticité*, Polytech. Paris, PST No. 116, pp. 67–113.
[37] J. Mandel (1966) Sur les équations d'écoulement des sols idéaux en déformation plane et le concept de double glissement, *J. Mech. Phys. Sol.*, **14**, 6, pp. 303–308.
[38] J. Mandel (1969) *Cours de Science des Matériaux*, Ecole National Supérieure des Mines de Paris.
[39] J. Mandel and F. Parsy (1961) Quelques problèmes tridimensionnel de la théorie du corps parfaitement plastique, *Sem. Plast. Ec. Polytechnique*, PST No. 116, pp. 105–127.
[40] J. Mandel and J. Salençon (1972) Force portante d'une fondation sol une assise rigide (étude théorique), *Géotechnique*, **22**, No. 1, pp. 79–93.
[41] J. Massau (1899) Mêmoire sur l'intégration graphique des equations aux dérivés partielles. Chap. IV. Equilibre limite des terres cohésion. repr. as *Edition du Centenaire*, Comité National de Méce Bruxelles, 1952, from *Ann. Ass. Ing. Ec. Gand.*
[42] G. Meyerhof (1950)–1951) The ultimate bearing capacity of foundation, *Géotechnique*, **2**, No. 4, pp. 301–332.
[43] A. Nadai (1950–1963) *Theory of Flow and Fracture of Solids*, Vol. II, McGraw Hill Book Co., New York.
[44] R. Negre (1968) Contribution à l'étude de l'équilibre limite des sols et des matériaux pulvérulents et cohérents, Thesis Doct. Sc. Grenoble.

[45] Nguyen Chanh (1968) Etude expérimentale de la poussée et de la butée des terres. *Ann. des Ponts et Chaussées,* No. 3, 1968, pp. 225–236.
[46] J. P. Obin (1972) Force portante en déformation plane d'un sol verticalement non homogène, Thesis, Univ. Grenoble.
[47] L. Prandtl (1933) Anwendungsbeispiele zu einem Henchyschen Satz über das plastische Gleichgewicht, *Z. Angew. Math. Mech.,* 3, p. 401.
[48] D. Radenkovic (1961) Théoremes limites pour un matériau de Coulomb à dilatation non standardisée, *C.R. Ac. Sc., Paris,* **252**, pp. 4103–4104.
[49] H. Reissner (1925) *Proc. 1st Intt Conf. Appl. Mech.*, Delft, p. 29
[50] C. Roche (1973) Force portante d'une fondation superficielle. *Trans. fin d'études E.N.P.C.,* Lab. Mécanique des Solides, Ec. Polytech. Paris.
[51] J. Salençon (1972) Butée d'une paroi lisse sur un massif plastique solutions statiques, *J. Mécanique,* **11**, No. 1, pp. 135–146.
[52] J. Salençon (1972) Prolongement des champs de Prandtl dans le cas du matériau de Coulomb, *Archives of Mechanics,* **25**, No. 4, pp. 643–648.
[53] J. Salençon (1974) *Plasticité pour la Mécanique des Sols,* C.I.S.M., Rankine Session, July 1974, Udine, Italy.
[54] R. T. Shield (1955) On the plastic flow of metals under conditions axial symmetry, *Proc. Roy. Soc.,* **233**, A, nr. 1193, pp. 267–287.
[55] V. V. Sokolovski (1960) *Statics of Soil Media,* Butterworths, London.
[56] V. V. Sokolovski (1962) Complete plane problems of plastic flow, *Jl. Mech. Phys. Solids,* **10**, pp. 353–364.
[57] V. V. Sokolovski (1963) Limit equilibrium of granular medium with variable weight, *Jl. Mech. Phys. Solids,* **11**, No. 6, pp. 395–410.
[58] V. V. Sokolovski (1965) *Statics of granular media,* Pergamon Press, Oxford.
[59] A. J. M. Spencer (1964) A theory of the kinematics of ideal soils in plane strain conditions, *J. Mech. Phys. Sol.,* **12**, 5, pp. 437–351.
[60] W. Szczepinski (1967) Wstep do analisy procesow obrobki, *Inst. Podst. Probl. Techn. Polish Ac. Sc.*
[61] J. P. Tournier (1972) Compartement d'une couche compressible limit par un substratum rigide et soumise à une change verticale applique par une semelle filante, Thesis Ph. D., Sherbrooke Univ., Canada.
[62] G. P. Tschebotarioff and J. R. Bayliss (1948) Determination of the shearing strength of varved clays and their sensibility to revolding, *Proc. 2nd Int. Conf. Soil Mech. Rotterdam,* **1**, pp. 203–207.
[63] A. Winzer and G. F. Carrier (1949) Discontinuities of stress in plane plastic flow, *Jl. Appl. Mech., Trans. A.S.M.E.,* **16**, pp. 346–348.

CHAPTER IV

Appendixes

A. PROBLEMS OF UNCONTAINED PLASTIC FLOW IN PLANE STRAIN FOR ISOTROPIC NON-HOMOGENEOUS MATERIAL

1. General

The problem of uncontained flow in plane strain is considered for an isotropic, non-homogeneous, rigid-plastic material, assuming that there is homogeneity normal to the plane of deformation.

As in the case of homogeneity, it is shown that if the material is 'standard' (i.e. it behaves according to the principle of maximum plastic work), any yield criterion may be reduced, for plane strain problems, to an intrinsic curve criterion in the plane of the strain.

Alternatively, if the criterion is of the intrinsic curve type, it suffices that the plane of straining contains the principal stresses so that the preceding conclusion still holds. This condition can result from relatively broad assumptions for the flow rule.

Further, the problems of plane strain, for the case of materials having an intrinsic curve criterion in the plane of the strain, i.e. a criterion whose expression varies with the location, will be studied.

2. The Problem of the Stresses—General Case

2.1 Stress characteristics

For the stresses in the plastic zone the system of equations is

$$f(\sigma_x, \sigma_y, \tau_{xy}, x, y) = 0 \tag{1}$$

$$\frac{\partial \sigma_x}{\partial x} + \frac{\partial \tau_{xy}}{\partial y} + \rho X = 0 \tag{2}$$

$$\frac{\partial \tau_{xy}}{\partial x} + \frac{\partial \sigma_y}{\partial y} + \rho Y = 0 \tag{3}$$

in which the criterion f explicitly depends on point (x, y).[1] The existence of continuous first derivatives of f with respect to x and y is assumed.

On introducing the parameters R, p, θ and ordering the principal stresses in the (x, y) plane according to $\sigma_1 \geqslant \sigma_2$, $\sigma_x, \sigma_y, \tau_{xy}$ are given by

$$\left.\begin{array}{rl} \sigma_x &= -p + R\cos 2\theta \\ \sigma_y &= -p - R\cos 2\theta \\ \tau_{xy} &= R \sin 2\theta \end{array}\right\}$$

$$(p = -(\sigma_1 + \sigma_2)/2,\ R = (\sigma_1 - \sigma_2)/2,\ \theta = (Ox, \sigma_1)) \tag{4}$$

The yield criterion, which is an intrinsic curve in the (x, y) plane, is solved as follows.

$$R = R(p, x, y) \tag{5}$$

and the angle ϕ, defined in Figure IV.A.1, for the intrinsic curve at point (x, y) makes it possible to evaluate the derivative:

$$\frac{\partial R}{\partial p} = \sin \phi \tag{6}$$

Equation (1) is equivalent to (4) and (5); and (2) and (3) are transformed into

$$-\frac{\partial p}{\partial x}(1 - \sin\phi \cos 2\theta) - 2R\sin 2\theta \frac{\partial \theta}{\partial x} + \sin\phi \frac{\partial p}{\partial y}\sin 2\theta + 2R\cos 2\theta \frac{\partial \theta}{\partial y}$$

$$+ \frac{\partial R}{\partial x}\cos 2\theta + \frac{\partial R}{\partial y}\sin 2\theta + \rho X = 0 \tag{7}$$

$$+ \frac{\partial p}{\partial x}\sin\phi \sin 2\theta + 2R\cos 2\theta \frac{\partial \theta}{\partial x} - \frac{\partial p}{\partial y}(1 + \sin\phi \cos 2\theta) + 2R\sin 2\theta \frac{\partial \theta}{\partial y}$$

$$+ \frac{\partial R}{\partial x}\sin 2\theta - \frac{\partial R}{\partial y}\cos 2\theta + \rho Y = 0 \tag{8}$$

$R, \phi, \partial R/\partial x, \partial R/\partial y$ are known functions of p, x and y. p and θ are unknown functions of both variables x and y.

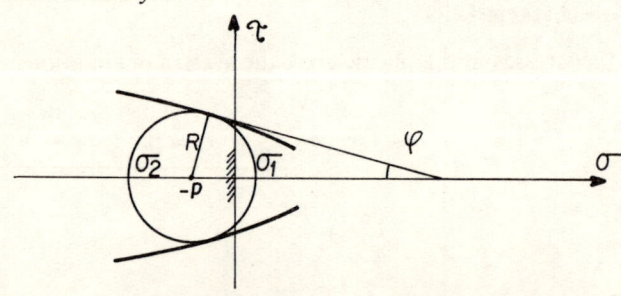

Figure IV.A.1

---

[1] (x, y) is a system of orthogonal coordinates. X and Y are the components of the body force $\mathbf{F}$ along the axes.

The system (7, 8) is of the same type, and has identical properties to that obtained in the homogeneous case.

The slopes of the characteristic lines are once again

$$dy/dx = \tan\left[\theta \pm \left(\frac{\pi}{4} + \frac{\phi}{2}\right)\right] \tag{9}$$

and these directions, which correspond to the points of contact of the Mohr circle with the intrinsic curve at point (x, y) are denoted by β (resp. α). These are the surfaces on which

$$|\tau| = h(\sigma, x, y) \tag{10}$$

(the equation of the intrinsic curve).

The relations along the characteristics are obtained by means of a standard method. For example, by siting the axes x and y along $M\alpha$ and its normal and combining (7) and (8) the equation for line α becomes

$$-\frac{\partial p}{\partial x}\cos\phi - 2R\frac{\partial \theta}{\partial x} + \frac{\partial R}{\partial y} + \rho(X\cos\phi + Y\sin\phi) = 0 \tag{11}$$

in which only the derivatives with respect to x of the unknown functions p and θ play a role ($\partial R/\partial y$ being known).

2.2 Coordinate system associated with the characteristic lines

X^α and X^β form a new system of coordinates, the coordinate lines of which are the α and β lines (Fiugre IV.A.2), and E_α, E_β at each point M are the local vectors linked to this system:

$$E_\alpha = \frac{\partial M}{\partial X^\alpha}, \quad E_\beta = \frac{\partial M}{\partial X^\beta}$$

A vector at M is written as $\mathbf{dM} = dX^\alpha E_\alpha$

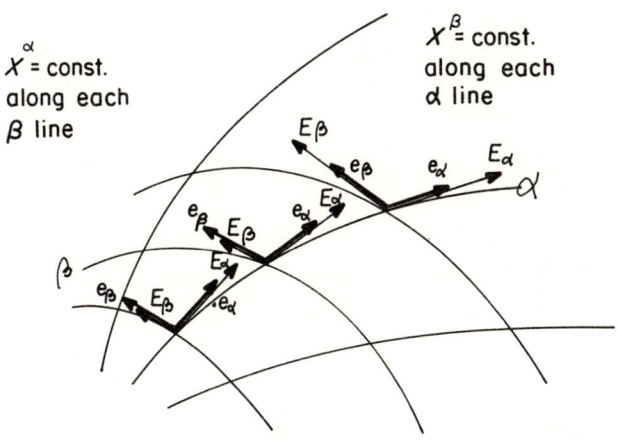

Figure IV.A.2

Then, the associated unit basis is given by

$$e_\alpha = E_\alpha/|E_\alpha|, \quad e_\beta = E_\beta/|E_\beta|$$

and $dM = dx^\alpha e_\alpha$. (The x^α are non-holonomial coordinates.) e^α, e^β form the dual basis (associated by the Euclidian inner product) and

$$dM = dx_\alpha \cdot e^\alpha.\text{[1]}$$

The following notation is used.

$$\partial_\alpha = \frac{\partial}{\partial x^\alpha}, \quad \partial^\alpha = \frac{\partial}{\partial x_\alpha}$$

2.3 Relations along the characteristics

Equation (11) may now be written as

$$\partial_\alpha p + \frac{2R}{\cos\phi} \partial_\alpha \theta - [\rho F^\alpha + \partial^\alpha R] = 0 \tag{12}$$

whilst

$$dp + \frac{2R}{\cos\phi} d\theta - \left(\rho F^\alpha + \frac{\partial R}{\partial x_\beta}\right) dx^\alpha = 0.\text{[2]} \tag{13}$$

is the relation along the α characteristic. Similarly, along a β characteristic,

$$dp - \frac{2R}{\cos\phi} d\theta - \left[\rho F^\beta + \frac{\partial R}{\partial x_\alpha}\right] dx^\beta = 0 \tag{14}$$

These equations are, for a non-homogeneous material with any intrinsic curve, the homologues of Kötter's equations for a homogeneous Coulomb material.

It can be seen that these equations are deduced from those obtained in the homogeneous case, by replacing $R(p)$ and $\phi(p)$ by $R(p, x, y)$ and $\phi(p, x, y)$ and by adding fictitious body forces.[3]

---

[1] The contravariant coordinates dx^α, dx^β are the oblique components of dM on the unit basis, tangential to the lines α, β; dx_α, dx_β are the orthogonal projections of this very vector on this basis.

[2] The oblique components along M_α, M_β of a vector with X, Y coordinates following any orthogonal axes are

$$F^\alpha = \frac{1}{\cos\phi}\left[X\sin\left\{\theta + \left(\frac{\pi}{4} + \frac{\phi}{2}\right)\right\} - Y\cos\left\{\theta + \left(\frac{\pi}{4} + \frac{\phi}{2}\right)\right\}\right]$$

$$F^\beta = \frac{1}{\cos\phi}\left[-X\sin\left\{\theta - \left(\frac{\pi}{4} + \frac{\phi}{2}\right)\right\} + Y\cos\left\{\theta - \left(\frac{\pi}{4} + \frac{\phi}{2}\right)\right\}\right]$$

whence the formulae giving $\partial R/\partial x_\alpha = \partial^\alpha R$ and $\partial R/\partial x_\beta = \partial^\beta R$ as functions of $\partial R/\partial x$ and $\partial R/\partial y$.

[3] Attention must be paid to the inversion of sugscripts: the fictitious body forces are not derived using R as a potential.

3. The Case of a Tresca Material

In the case of a Tresca material with a variable shear strength $k(x, y)$ (see also [18]),

$$\partial^\alpha R = \frac{\partial k}{\partial x_\alpha} = \frac{\partial k}{\partial x} \cos\left(\theta - \frac{\pi}{4}\right) + \frac{\partial k}{\partial y} \sin\left(\theta - \frac{\pi}{4}\right)$$

$$\partial^\beta R = \frac{\partial k}{\partial x_\beta} = -\frac{\partial k}{\partial x} \sin\left(\theta - \frac{\pi}{4}\right) + \frac{\partial k}{\partial y} \cos\left(\theta - \frac{\pi}{4}\right)$$

α and β lines are orthogonal, and $dx^\alpha = dx_\alpha$, $F^\alpha = F_\alpha$, etc. Equations (13) and (14) are written

$$dp + 2k(x, y)\, d\theta - \left(\frac{\partial k}{\partial x_\beta} + \rho F^\alpha\right) dx^\alpha = 0 \quad \text{for an } \alpha \text{ line}$$

$$dp - 2k(x, y)\, d\theta - \left(\frac{\partial k}{\partial x_\alpha} + \rho F^\beta\right) dx^\beta = 0 \quad \text{for a } \beta \text{ line}$$

Assuming the body forces are derived from a potential V, $F^\alpha = F_\alpha = -\partial V/\partial x^\alpha$ and

$$\left.\begin{array}{l} d(p + \rho V) + 2k\, d\theta - \dfrac{\partial k}{\partial x_\beta} dx^\alpha = 0, \quad \text{for an } \alpha \text{ line} \\[1em] d(p + \rho V) - 2k\, d\theta - \dfrac{\partial k}{\partial x_\alpha} dx^\beta = 0, \quad \text{for a } \beta \text{ line} \end{array}\right\} \quad (15)$$

4. The Case of a Coulomb Material

For a non-homogeneous Coulomb material,

$$R(p, x, y) = \rho \sin[\phi(x, y)] + C(x, y) \cdot \cos[\phi(x, y)]$$

whence,

$$\partial R = \cos\phi \cdot \partial C + (p\cos\phi - C\sin\phi)\, \partial\phi$$

In particular, if the material has a constant internal angle of friction and a variable cohesion, the network of characteristics is isogonal and relations (13) and (14) are written

$$dp + \frac{2R}{\cos\phi} d\theta - \left(\rho F^\alpha + \cos\phi \frac{\partial C}{\partial x_\beta}\right) dx^\alpha = 0, \quad \text{for an } \alpha \text{ line}$$

$$dp - \frac{2R}{\cos\phi} d\theta - \left(\rho F^\beta + \cos\phi \frac{\partial C}{\partial x_\alpha}\right) dx^\beta = 0, \quad \text{for a } \beta \text{ line}$$

5. The Velocity Problem

5.1 Flow rule

The stress problem can be solved as indicated without having the flow rule fully defined, provided that it satisfies certain conditions. In order to solve the velocity problem it is necessary to have the flow rule completely defined. The case of a standard material will be considered, and then it will be seen how to deal with the case of some non-standard materials.

Assuming the stress problem is solved within the general framework of Section 2, the velocity problem may be formulated as follows. The two-dimensional yield criterion in the plane of the deformation also appears as the two-dimensional plastic potential. Therefore,

$$\left.\begin{aligned} v_{11} &= \lambda \frac{\partial f}{\partial \sigma_1} \\ v_{12} &= v_{21} = 0 \\ v_{22} &= \lambda \frac{\partial f}{\partial \sigma_2} \\ \lambda &\geq 0 \end{aligned}\right\} \quad (16)$$

The expression of f as a function of the principal stresses is easily obtained: in fact, $f = R - R(p)$ with the definitions of Section 2.1., i.e.,

$$f(\sigma_1, \sigma_2) \equiv \frac{\sigma_1 - \sigma_2}{2} - R\left(-\frac{\sigma_1 + \sigma_2}{2}\right) \quad (17)$$

whence, according to (6),

$$\left.\begin{aligned} v_{11} &= \lambda(1 + \sin \phi) \\ v_{12} &= v_{21} = 0 \\ v_{22} &= -\lambda(1 - \sin \phi) \\ \lambda &\geq 0 \end{aligned}\right\} \quad (18)$$

The value of ϕ is known at each point (x, y) from the previous stress solution and the velocity problem is seen to be linear.

5.2 Velocity characteristics

Again the coordinates x^α, x^β, defined in Section 2.2, are used. The components of tensor v_{ij} are

$$\left.\begin{array}{l}v_{\alpha\alpha} = \lambda\left[(1 + \sin\phi)\cos^2\left(\frac{\pi}{4} + \frac{\phi}{2}\right) - (1 - \sin\phi)\sin^2\left(\frac{\pi}{4} + \frac{\phi}{2}\right)\right] = 0 \\ v_{\alpha\beta} = v_{\beta\alpha} = \lambda\left[(1 + \sin\phi)\cos^2\left(\frac{\pi}{4} + \frac{\phi}{2}\right)\right. \\ \left. + (1 - \sin\phi)\sin^2\left(\frac{\pi}{4} + \frac{\phi}{2}\right)\right] = \lambda\cos^2\phi \\ v_{\beta\beta} = \lambda\left[(1 + \sin\phi)\cos^2\left(\frac{\pi}{4} + \frac{\phi}{2}\right) - (1 - \sin\phi)\sin^2\left(\frac{\pi}{4} + \frac{\phi}{2}\right)\right] = 0 \\ \lambda \geqslant 0 \end{array}\right\} \quad (19)$$

or, alternatively,

$$\left.\begin{array}{l} D_\alpha v_\alpha = 0 \\ D_\beta v_\beta = 0 \\ \tfrac{1}{2}(D_\alpha v_\beta + D_\beta v_\alpha) \geqslant 0 \end{array}\right\} \quad (20)$$

where the D's denote the covariant derivatives, the expressions of which are as follows.

$$\left.\begin{array}{l} D_\alpha v_\alpha = \partial_\alpha v_\alpha - v_\alpha \tan\phi\, \partial_\alpha\left(\theta - \frac{\phi}{2}\right) - v_\beta \frac{1}{\cos\phi} \partial_\alpha\left(\theta - \frac{\phi}{2}\right) \\ D_\beta v_\alpha = \partial_\beta v_\alpha - v_\alpha \tan\phi\, \partial_\beta\left(\theta - \frac{\phi}{2}\right) - v_\beta \frac{1}{\cos\phi} \partial_\beta\left(\theta - \frac{\phi}{2}\right) \\ D_\alpha v_\beta = \partial_\alpha v_\beta + v_\alpha \frac{1}{\cos\phi} \partial_\alpha\left(\theta + \frac{\phi}{2}\right) + v_\beta \tan\phi\, \partial_\alpha\left(\theta + \frac{\phi}{2}\right) \\ D_\beta v_\beta = \partial_\beta v_\beta + v_\alpha \frac{1}{\cos\phi} \partial_\beta\left(\theta + \frac{\phi}{2}\right) + v_\beta \tan\phi\, \partial_\beta\left(\theta + \frac{\phi}{2}\right) \end{array}\right\} \quad (21)$$

$\partial_\alpha \theta$ and $\partial_\beta \theta$ are known (from the stress solution), as are $\partial_\alpha \phi$ and $\partial_\beta \phi$ which are decomposed into two terms:[1]

$$\partial_\alpha \phi = \partial_\alpha p \cdot \frac{\partial\phi}{\partial p} + \frac{\partial\phi}{\partial x} a_\alpha^i \quad (22)$$

(the term of the intrinsic curve plus the term of non-homogeneity).

---

[1] x^i are fixed orthogonal coordinates in which the non-homogeneity is defined. The a_α^i's denote the cosines of the axes x^i with respect to the tangents to the α and β lines at the considered point:

$$a_\alpha^1 = \cos\left(\theta - \frac{\pi}{4} - \frac{\phi}{2}\right) \qquad a_\alpha^2 = \sin\left(\theta - \frac{\pi}{4} - \frac{\phi}{2}\right)$$

$$a_\beta^1 = \cos\left(\theta + \frac{\pi}{4} + \frac{\phi}{2}\right) \qquad a_\beta^2 = \sin\left(\sigma + \frac{\pi}{4} + \frac{\phi}{2}\right)$$

5.3 Interpretation of the results

$D_\alpha v_\alpha = 0$ implies that the extension rate along the α line is zero. Therefore, the α line is a characteristic for the velocities and

$$D_\alpha v_\alpha = \partial_\alpha v_\alpha - v_\alpha \tan \phi \partial_\alpha \left(\theta - \frac{\phi}{2}\right) - v_\beta \frac{1}{\cos \phi} \partial_\alpha \left(\theta - \frac{\phi}{2}\right) = 0 \quad (23)$$

is the relation along this characteristic. The same applies for the β line, with

$$D_\beta v_\beta = \partial_\beta v_\beta + v_\alpha \frac{1}{\cos \phi} \partial_\beta \left(\theta + \frac{\phi}{2}\right) + v_\beta \tan \phi \partial_\beta \left(\theta + \frac{\phi}{2}\right) = 0 \quad (24)$$

Finally,

$$D_\alpha v_\beta + D_\beta v_\alpha = \partial_\beta v_\alpha + \partial_\alpha v_\beta - \left(v_\alpha \tan \phi + v_\beta \frac{1}{\cos \phi}\right) \partial_\beta \left(\theta - \frac{\phi}{2}\right)$$

$$+ \left(v_\alpha \frac{1}{\cos \phi} + v_\beta \tan \phi\right) \partial_\alpha \left(\theta + \frac{\phi}{2}\right) \geq 0 \quad (25)$$

is the condition of positivity (of λ).

In the case of a Tresca criterion, the Geiringer relations are recognized in equations (23, 24), also in (25), the condition of positivity for the homogeneous case (the non-homogeneity having no influence on the form of the relations in this case).

As a rule, the determination of the velocity field in the plastic zones is effected using the method of characteristics (relations (23) and (24), the condition (25) then being verified at each point).

5.4 Hodograph

It is convenient to express these results by introducing the velocity plane (a method developed by Green [13] for the case of a homogeneous Tresca material).

Relations (23, 24) show that at one point, an α (resp. β) line and its image a (resp. b) on the hodograph (the a, b network being the image of the α, β network, in the plane u_x, u_y) are orthogonal. Therefore, the situation is represented by one of the four configurations of Figure IV.A.3.

If $M + \mathbf{dM}$ is a point of the first sector (α, β), i.e.

$$dx^\alpha > 0 \quad \text{and} \quad dx^\beta > 0$$

the condition of positivity (25) then gives

$$(D_\alpha v_\beta + D_\beta v_\alpha) dx^\alpha dx^\beta \geq 0 \quad (26)$$

Also, $D_\alpha v_\beta dx^\alpha = Dv_\beta$, (24) being taken into account and $D_\beta v_\alpha dx^\beta = Dv_\alpha$, (23) being taken into account. Therefore, (26) may be written,

$$Dv_\alpha dx^\alpha + Dv_\beta dx^\beta = D\mathbf{u}(\mathbf{dM}) \cdot \mathbf{dM} \geq 0 \quad (27)$$

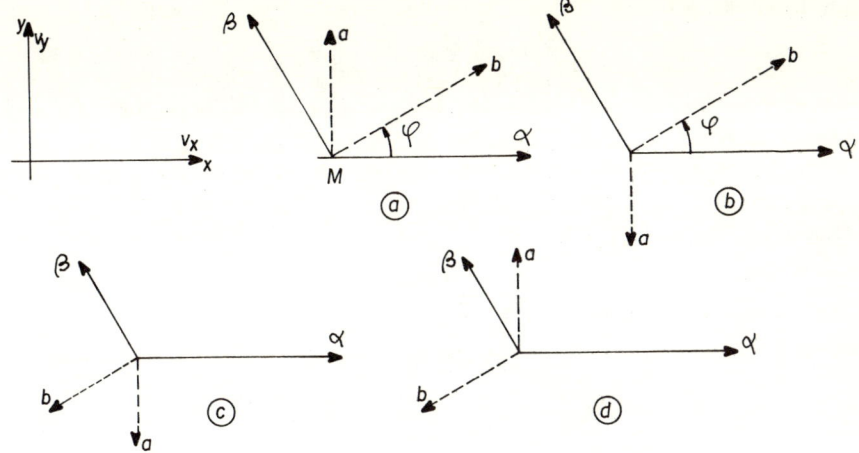

Figure IV.A.3

and the reciprocal can easily be seen, i.e. if (27) is satisfied $\forall d\mathbf{M}$ in the sector (α, β) then (25) is verified.

Obviously, if configuration (a) is relevant (27) is certainly verified, but for (c), (27) is not verified. There is some doubt concerning configurations (b) and (d).

The construction of the hodograph often provides a convenient means to check if the condition of positivity is fulfilled (a result indicated by Ewing and Hill [7] in the case of a homogeneous Tresca material).

5.5 Discontinuity of the velocity

As the velocity problem is linear hyperbolic, the discontinuity lines of the velocity are the characteristics of the problem, i.e. the α and β lines.

Writing relations (23–25) in terms of discontinuities gives for (23), when crossing a β line (i.e. following an α line),

$$[v_\alpha] = 0$$

and the discontinuity concerns v_β only; and for (24), when crossing an α line,

$$[v_\beta] = 0$$

and the discontinuity concerns v_α only. These are the discontinuity conditions.

The propagation equations of the discontinuity are obtained by applying (23) or (24) on both sides of the corresponding discontinuity line. Thus, following an α line,

$$D_\alpha[v_\alpha] = \partial_\alpha[v_\alpha] - [v_\alpha]\tan\phi\partial_\alpha\left(\theta - \frac{\phi}{2}\right) = 0$$

and following a β line,

$$D_\beta[v_\beta] = \partial_\beta[v_\beta] + [v_\beta]\tan\phi\partial_\beta\left(\theta + \frac{\phi}{2}\right) = 0$$

equations which are easily integrated if $\phi =$ constant.

Finally, from (25), come the conditions of positivity. On crossing a β line in the direction of an α line, $[v_\beta] \geq 0$ and likewise, on crossing an α line in the direction of a β line, $[v_\alpha] \geq 0$.

Summary:

crossing of α line: $\quad [v_\beta] = 0$

$\qquad\qquad\qquad\qquad [v_\alpha] \geq 0, D_\alpha[v_\alpha] = 0$

if $\phi =$ constant $[v_\alpha] = [v_\alpha]_0 \exp[(\theta - \theta_0)\tan\phi]$

crossing of β line: $\quad [v_\alpha] = 0$

$\qquad\qquad\qquad\qquad [v_\beta] \geq 0, D_\beta[v_\beta] = 0$

if $\phi =$ Constant $[v_\beta] = [v_\beta]_0 \exp[-(\theta - \theta_0)\tan\phi]$

On the hodograph, the vectors representing the discontinuity are necessarily oriented according to M_a or M_b from the (a) configuration. It is to be emphasized that only in the case of a Treasca material is the velocity discontinuity a tangent to the discontinuity lines.

5.6 A comment on the interpretation of expressions (23), (24), (25)

In order to prepare for the study of a non-standard material, a comment is made on the interpretation of (23), (24), (25). The equations (23) and (24) imply that the strain rate in the plane allows zero extension along the α and β lines which form the angle $(\pi/2 + \phi)$ and are bisected by the principal directions of $\boldsymbol{\sigma}$. Therefore, the strain rate $\mathbf{v}$ has the same principal directions as $\boldsymbol{\sigma}$, and satisfies the flow rule

$$v_1 = \lambda(1 + \sin\phi)$$
$$v_2 = -\lambda(1 - \sin\phi)$$

and (24) indicates that λ is positive.

6. The Case of Some Non-standard Materials

A non-homogeneous, non-standard material is now considered, having a yield criterion which assumes the intrinsic curve form,

$$R = R(p, x, y) \tag{5}$$

whence,

$$\frac{\partial R}{\partial p} = \sin\phi\,(p, x, y) \tag{6}$$

The flow rule is such that **v** has the same principal directions as **σ** and

$$\left. \begin{aligned} v_1 &= \lambda(1 + \sin v(p, x, y)) \\ v_2 &= 0 \\ v_3 &= -\lambda(1 - \sin v(p, x, y)) \\ \lambda &\geq 0 \\ v(p, x, y) &\leq \phi(p, x, y) \\ (\sigma_1 &\geq \sigma_2 \geq \sigma_3) \end{aligned} \right\} \quad (28)$$

Therefore, these are non-standard materials with a yield criterion of the intrinsic curve form, and also a plastic potential of the intrinsic curve type, both functions depending on x and y due to lack of homogeneity. This is a generalization of a Coulomb material with a Coulomb or Tresca plastic potential, as studied by numerous authors (e.g. [2, 4–6, 14, 16, 17, 29, 31, 32]).

According to the flow rule the deformation plane contains the extreme principal stresses in the plastic zones. Thus, the problem for the stresses is set and solved as in Section 2.

As the stress field is determined in the plastic zones the velocity problem can also be solved in the same regions.

According to Section 5.6 the strain rate **v** at each point must admit as directions of zero extension the directions γ and δ, which form with σ_1 the known angle

$$\mp \frac{1}{2}\left(\frac{\pi}{2} + v(p, x, y)\right) \quad (29)$$

and the factor λ must also be positive (Figure IV.A.4). Expressed differently, the velocity problem, being linear and hyperbolic, admits as its characteristic lines the (γ, δ) lines, with slopes given by

$$\frac{dy}{dx} = \tan\left[\theta \mp \left(\frac{\pi}{2} + v\right)\right] \quad (30)$$

The relations along those characteristics are

$$D_\gamma v_\gamma = 0 \text{ along a } \gamma \text{ line} \quad (31)$$

$$D_\delta v_\delta = 0 \text{ along a } \delta \text{ line} \quad (32)$$

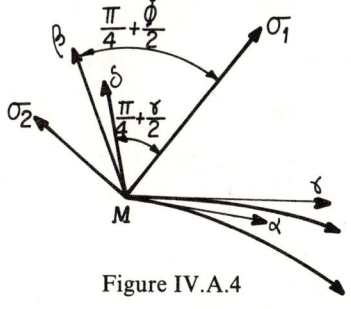

Figure IV.A.4

The condition of positivity of λ is

$$D_\gamma v_\delta + D_\delta v_\gamma \geq 0 \qquad (33)$$

Equations (31, 32, 33) are obtained from (23-25) by replacing (α, β, ϕ) by (γ, δ, ν).[1]

EXAMPLE: Taken as an example is the problem of the indentation of a half-plane of a homogeneous, weightless Coulomb material, with an internal angle of friction ϕ and zero angle of dilation ν (deformation without volume change).

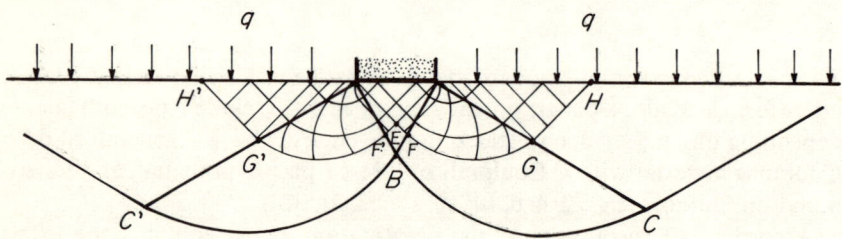

Figure IV.A.5

Figure IV.A.5 represents the networks of the characteristics (α, β) and (γ, δ). The (α, β) characteristic network is classical, consisting of homogeneous fields with rectilinear characteristics, and Prandtl's fan (vector radii and logarithmic spirals). The (γ, δ) characteristic network consists of orthogonal lines: straight lines in $AFEF'A'$, AGH, $A'G'H'$, and logarithmic spirals in AFG and $A'F'G'$.

The pressure on $A'A$ is given by the usual result

$$P_{\text{ult}} = -H + (q + H)\frac{1 + \sin\phi}{1 - \sin\phi}e^{\pi \tan\phi}$$

The velocity field is obtained by integrating the Geiringer equations numerically (or constructing the hodograph) in $A'EFGHA$. It is symmetrical.

The same problem has been dealt with in [2] for a Coulomb material, taking weight and surface load into account, with various assumptions on the value of the constant angle ν. The construction of the characteristic networks (α, β) and (γ, δ) is more difficult than before, as the determination of the four families of lines must be carried out jointly. The construction firstly utilizes the characteristics (α, β) but underneath the footing it requires the participation of the (γ, δ) characteristics. According to the numerical results obtained in (2) the value of ν seems to exert little influence on the bearing capacity (for the case of a weightless material, it has no influence, since the stress field is determined independently of ν).

The checking of the condition of positively of factor λ (of relation (28)), is carried out numerically. In [2], Davis and Booker, having carried out calculations for various values of ϕ and ν, state that the dissipation is always positive.

---

[1] For the utilization of the hodograph (Section 5.5), $(\alpha, \beta, \phi, a, b)$ must be replaced by $(\gamma, \delta, \nu, c, d)$.

However, Drescher [4, 5], on dealing with the example of Figure IV.A.5, which is a particular case of the problem studied in (2), shows that the dissipation is negative in some zones when $v \neq \phi$.[1]

7. Discontinuity of the Stress-field

The solutions considered in Section 2 for the stress problem are only applicable where the stress-field is continuous. In some problems, weak solutions for stresses must be considered, i.e. solutions admitting lines of discontinuity of the stress-field.

These solutions will be examined firstly from the viewpoint of stresses, and then of the strain rates, in order to determine the necessary conditions for the velocity field along a line of discontinuity of stresses.

7.1 Conditions for stresses

The stress problem as defined by (7. 8) is hyperbolic and quasi-linear. It is known that

(1). The discontinuity lines of the weak solutions for such a problem are not the characteristic lines;

(2). Weak solutions can be obtained even from continuous data;

(3). Discontinuity conditions exist, but are not sufficient for the determination of a weak solution. Recourse must be made to a supplementary condition in order to determine the discontinuity.

Discontinuity conditions:

The discontinuity conditions for the stress-field are obtained by applying the continuity of stress to the surface of discontinuity (Figures IV.A. 6a and b). With the axis Ox placed, for convenience, along the normal to the surface of discontinuity at M,

$$-p_1 + R(p_1, x, y) \cos 2\theta_1 = -p_2 + R(p_2, x, y) \cos 2\theta_2$$

$$R(p_1, x, y) \sin 2\theta_1 = R(p_2, x, y) \sin 2\theta_2$$

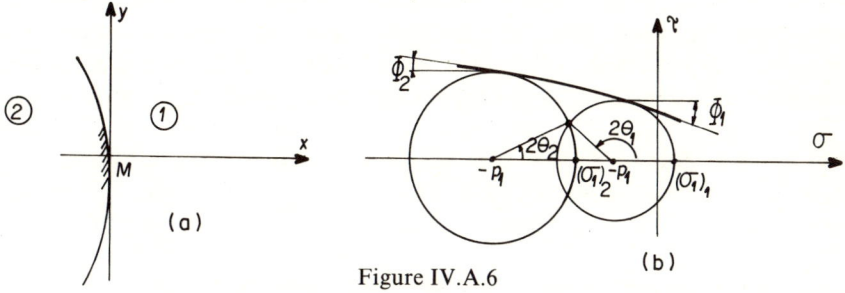

Figure IV.A.6

[1] For $v = 0$, there is equivalence between the condition $\lambda \geq 0$ (generally more restrictive) and the thermodynamic condition of non-negativity of the dissipation.

In the particular case of a Coulomb material (ϕ independent of p), equation (34) leads to

$$\sin(\theta_1 - \theta_2)[\sin\phi \cos(\theta_1 - \theta_2) - \cos(\theta_1 + \theta_2)] = 0 \tag{35}$$

whence, if the solution corresponding to the continuous field is ignored, the relation between θ_1 and θ_2 in the case of the discontinuity is

$$\sin\phi \cos(\theta_1 - \theta_2) - \cos(\theta_1 + \theta_2) = 0 \tag{36}$$

and therefore,

$$\tan\theta_2 = \cot\theta_1 \frac{1 - \sin\phi}{1 + \sin\phi}$$

Figure IV.A. 6*b* clearly shows that in the case of any intrinsic curve (convex and 'opened' in the direction $\sigma < 0$) the values of θ_1 and θ_2 corresponding to the discontinuity are situated, with respect to the values

$$\phi_1 = \phi(p_1, x, y) \quad \text{and} \quad \phi_2 = \phi(p_2, x, y) \tag{37}$$

as indicated in the following table (38).

θ_1	$\frac{\pi}{2} > \theta_1 > \frac{\pi}{4} - \frac{\phi_1}{2}$		$\frac{\pi}{4} - \frac{\phi_1}{2} > \theta_1 > -\frac{\pi}{4} + \frac{\phi_1}{2}$		$-\frac{\pi}{4} + \frac{\phi_1}{2} > \theta_1 > -\frac{\pi}{2}$
	$\frac{\pi}{2}$	↘	$\frac{\pi}{4} - \frac{\phi}{2}$	↘	0 ↘ $-\frac{\pi}{4} + \frac{\phi}{2}$ ↘ $-\frac{\pi}{2}$
θ_2	0	↗	$\frac{\pi}{4} - \frac{\phi}{2}$	↗	$\pi/2$ ↗ $\frac{3\pi}{4} + \frac{\phi}{2}$ ↗ π
	$0 < \theta_2 < \frac{\pi}{4} - \frac{\phi_2}{2}$		$\frac{\pi}{4} - \frac{\phi_2}{2} < \theta_2 < \frac{3\pi}{4} + \frac{\phi_2}{2}$		$\frac{3\pi}{4} + \frac{\phi_2}{2} < \theta_2 < \pi$

(38)

7.2 Interpretation of the line of discontinuity for the stresses

The line of discontinuity of the stress field may be considered as the limit of an infinitely thin transition zone which serves to connect two continuous limit equilibrium stress fields-producing opposite stresses on the faces of the zone (Figure IV.A.7*a*). In this zone the stress-field cannot be in limit equilibrium everywhere: if this were the case, a family of stress characteristics would necessarily have an envelope in this zone as in Figure IV.A.7*a*, and according to Bonneau's theorem, such a field cannot exist (see Appendix B of Chapter V.)

The transition zone, therefore, includes a part in which the criterion of limit equilibrium is not satisfied. This is (according to Hill [15]) a central core in which most of the transition occurs. The stresses τ_{xy} and σ_x vary little when crossing the transition zone and the Mohr circle representative of the stress state always passes closely to the point (σ, τ) corresponding to the stress applied to the faces of the zone in the states (1) and (2) (Figure IV.A.7*b*).[1]

---

[1] Moreover, the discontinuity lines of the stress field do appear thus in the solutions of some elastoplastic problems for a standard Tresca material, e.g. indentation of an acute angle wedge [20].

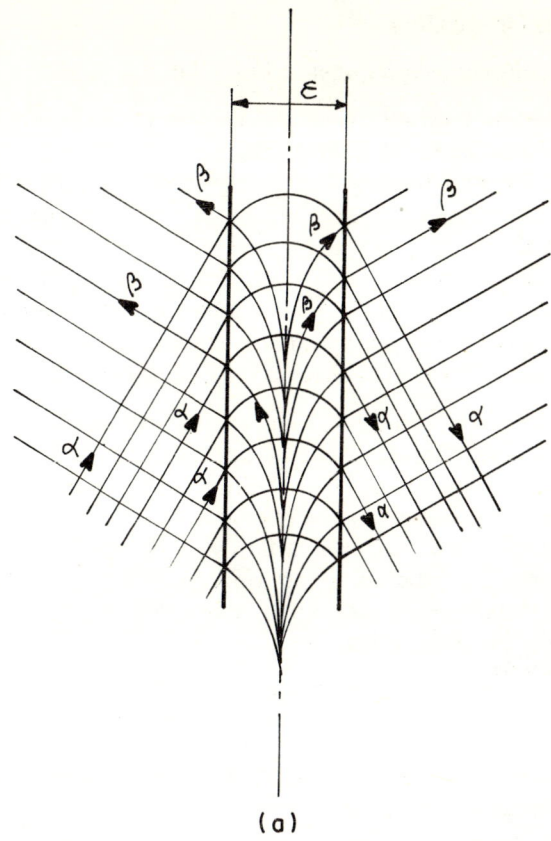

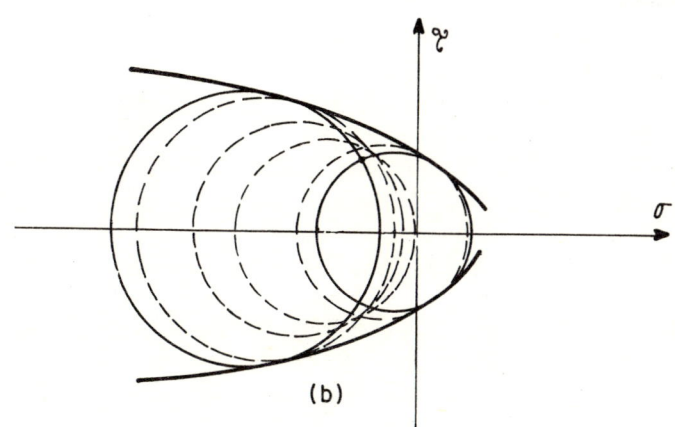

Figure IV.A.7

7.3 Conditions for velocities

7.3.1. *Discontinuity of velocity along a line (C) in a domain.*

As already state (Section 5), once the stress problem has been solved the velocity problem is linear and hyperbolic.

A classical theorem implies that, in each domain where the stress-field is continuous, the discontinuity lines for the velocities are the characteristics for the velocity problem.

If (C) is a line of discontinuity in the stress-field then, according to the previous statement, a discontinuity of the velocity along (C) in the domain (1) (resp. (2)) can exist only if (C) is a velocity characteristic in the domain.

Therefore, the following conclusions may be stated:

(1). *For a standard material*, it is known that the velocity characteristics are identical to the stress characteristics (Section 5.4.) and it follows from Section 7.1. that (C) is not a velocity characteristic. There can be no discontinuity of velocity along (C) either in domains (1) and (2).

(2). *For a non-standard material*, a flow rule of type (28) is adopted. As the material has any intrinsic curve (convex, open in the direction $\sigma > 0$) it is assumed that, beside the condition $v(p, x, y) \leq \phi(p, x, y)$, v is a non-increasing function of p. In these conditions, for

and
$$\left. \begin{array}{l} v_1 = v(p_1, x, y) \\ v_2 = v(p_2, x, y) \end{array} \right\} \quad (39)$$

it is possible to designate the values of θ_1 and θ_2 with respect to ϕ_1, ϕ_2, v_1 and v_2 as indicated in the following table (40).

θ_1	$\frac{\pi}{2} > \theta_1 > \frac{\pi}{4} - \frac{\phi_1}{2}$	$\frac{\pi}{4} - \frac{\phi_1}{2} > \theta_1 > -\frac{\pi}{4} + \frac{\phi_1}{2}$		$-\frac{\pi}{4} + \frac{\phi_1}{2} > \theta_1 > -\frac{\pi}{2}$		
	$\frac{\pi}{2} > \theta_1 > \frac{\pi}{4} - \frac{v_1}{2}$	$\frac{\pi}{4} - \frac{v_1}{2} > \theta_1 > -\frac{\pi}{4} + \frac{v_1}{2}$		$-\frac{\pi}{4} + \frac{v_1}{2} > \theta_1 > -\frac{\pi}{2}$		
	$\frac{\pi}{2}$ \ $\frac{\pi}{4} - \frac{v}{2}$ \ $\frac{\pi}{4} - \frac{\phi}{2}$ \	0 \	$-\frac{\pi}{4} + \frac{\phi}{2}$ \ $-\frac{\pi}{4} + \frac{v}{2}$ \	$-\frac{\pi}{2}$		
	0 ↗ $\frac{\pi}{4} - \frac{\phi}{2}$ ↗ $\frac{\pi}{4} - \frac{v}{2}$ ↗ $\frac{\pi}{2}$ ↗ $\frac{3\pi}{4} + \frac{v}{2}$ ↗ $\frac{3\pi}{4} + \frac{\phi}{2}$ ↗ π					
θ_2	$0 < \theta_2 < \frac{\pi}{4} - \frac{v_2}{2}$	$\frac{\pi}{4} - \frac{v_2}{2} < \theta_2 < \frac{3\pi}{4} + \frac{v_2}{2}$		$\frac{3\pi}{4} + \frac{v_2}{2} < \theta_2 < \pi$		
	$0 < \theta_2 < \frac{\pi}{4} - \frac{\phi_2}{2}$	$\frac{\pi}{4} - \frac{\phi_2}{4} < \theta_2 < \frac{3\pi}{4} + \frac{\phi_1}{2}$		$\frac{3\pi}{4} + \frac{\phi_2}{2} < \theta_2 < \pi$		

(40)

According to Section 6, the velocity characteristics are no longer identical to the stress characteristics. The line (C), without being a stress characteristic, can be a velocity characteristic in either domain (1) or (2). This occurs in domain (1) if

$$|\theta_1| = \frac{\pi}{4} - \frac{v_1}{2} \tag{41}$$

and in domain (2) if

$$\theta_2 = \frac{\pi}{4} - \frac{v_2}{2} \quad \text{or} \quad \theta_2 = \frac{3\pi}{4} + \frac{v_2}{2} \tag{42}$$

Table (40) shows that (41) and (42) cannot both be true simultaneously. Thus, in the case of a non-standard material of the type indicated, there can exist a discontinuity of the velocity along (C) in the domain (1) (resp. (2)), if condition (41) (resp. (42)) is satisfied along (C). Then (C) is a velocity characteristic in the domain (1) (resp. (2)) and the velocity discontinuity is governed by the equations of Section 5.6.

7.3.2. Discontinuity of the velocity from one domain to another, crossing (C).

By interpreting the discontinuity line for the stresses, as in Section 7.2, it can be seen whether this zone also makes a transition possible for the velocities, corresponding to a discontinuity from domain (1) to domain (2).

This transition can occur only in the plastic region of the zone, which according to the conclusions of Section 7.2 means that there is a discontinuity of the velocity along (C) in domain (1) or (2), as in the previous problem. Therefore, there is no discontinuity of the velocity from one domain to another, crossing a discontinuity line of the stress-field.

7.3.3. Inextensibility of the stress discontinuity line.

Since $[u] = 0$ and the velocity is continuous, the derivatives tangential to the stress discontinuity line are also continuous (Hadamard's relations) or,

$$[u_{x,y}] = 0$$
$$[u_{y,y}] = 0$$

Hence, for the strain-rate tensor

$$[v_{yy}] = 0 \tag{43}$$

Consequences:

A flow rule of type (28) is adopted, and on both sides of the discontinuity line

$$\left. \begin{array}{l} v_{yy} = (v_1) \sin^2 \theta + (v_2) \cos^2 \theta = \lambda[\sin v(p, x, y) - \cos 2\theta] \\ \lambda \geq 0 \end{array} \right\} \tag{44}$$

and (43) becomes

$$\left.\begin{array}{c}\lambda_1[\sin v(p_1, x, y) - \cos 2\theta_1] = \lambda_2[\sin v(p_2, x, y) - \cos 2\theta_2] \\ \lambda_1 \geq 0, \qquad \lambda_2 \geq 0\end{array}\right\} \quad (45)$$

According to the values of p_1 and p_2 (whence θ_1 and θ_2 by (34)), it may or may not be possible to determine values of $\lambda > 0$ such that (45) is satisfied. If this is not possible, the only solution will be $\lambda_1 = \lambda_2 = 0$ and hence,

$$(\mathbf{v})_1 = (\mathbf{v})_2 \quad (46)$$

The strain-rate tensors are zero on both sides of the stress discontinuity line, so that the line is inextensible.

Firstly, the case of a standard material with any (convex) intrinsic curve is examined. For each value of p,

$$v(p, x, y) = \phi(p, x, y)$$

It is possible to determine values of λ_1 and $\lambda_2 > 0$ if the solutions θ_1 and θ_2 of equation (34) corresponding to the discontinuity are such that

$$\rho(\theta_1, \theta_2) = \frac{\cos 2\theta_1 - \sin \phi_1}{\cos 2\theta_2 - \sin \phi_2} > 0 \quad (47)$$

It is immediately obvious from table (38) that

$$\rho(\theta_1, \theta_2) \leq 0$$

The only solution of (46) is, therefore,

$$\lambda_1 = \lambda_2 = 0$$

and hence

$$(\mathbf{v})_1 = (\mathbf{v})_2.$$

Thus, for a standard material with any (convex) intrinsic curve, the discontinuity line of stresses is inextensible.[1]

The case of a non-standard material of the type indicated in Section 7.3.1. is now considered. The sign of the ratio,

$$\rho(\theta_1, \theta_2, v_1, v_2) = \frac{\cos 2\theta_1 - \sin v_1}{\cos 2\theta_2 - \sin v_2} \quad (48)$$

is seen, by reference to table (40), to be not always negative. Therefore, in the case of a non-standard material, condition (45) is equivalent to the inextensibility of the discontinuity line only for some values of θ_1 and θ_2 among the solutions of (34).

---

[1] This result generalizes Geiringer's [12].

For a Coulomb material (ϕ and v independent of p) $\lambda_1, \lambda_2 \neq 0$ if the solutions θ_1 and θ_2 of equation (36) verify

$$\left. \begin{array}{c} 0 \leqslant |\theta_1| < \left(\dfrac{\pi}{4} - \dfrac{v}{2}\right) \\[2ex] \dfrac{\pi}{4} + \dfrac{v}{2} < \left|\dfrac{\pi}{2} - \theta_2\right| \leqslant \dfrac{\pi}{2} \end{array} \right\} \tag{49}$$

References

[1] D. Berthet, J. C. Hayot, and J. Salençon (1972) Poinçonnement d'un milieu semi-infini en matériau plastique de Tresca non-homogène, *Archives of Mechanics*, **24**, No. 1, pp. 127–138.
[2] E. H. Davis and J. R. Booker (1971) The bearing capacity of strip foot from the standpoint of plasticity theory, Univ. Sydney, Civ. Eng. Lab. Research rept. No. R 170.
[3] E. H. Davis and J. R. Booker (1973) The effect of increasing strength with depth on the bearing capacity of clays, *Géotechnique*, **23**, No. 4, pp. 551–563.
[4] A. Drescher (1971) A note on plane flow of granular media, *Probleme de la Rhéologie, Symp. Franco-Polonais*, Warsaw, 1971, pp. 135–144.
[5] A. Drescher (1972) Some remarks on plane flow of granular media, *Archives of Mechanics*, **24**, No. 5–6, pp. 837–848.
[6] A. Drescher, K. Kwaszczynska, and Z. Mroz (1967) Statics and kinematics of the granular medium in the case of wedge indentation, *Archives of Mechanics*, **19**, pp. 99–113.
[7] D. J. F. Ewing and R. Hill (1967) The plastic constraint of V. notched tension bars, *J. Mech. Phys. Solids*, **15**, No. 2, pp. 115–124.
[8] G. Favretti (1965) Impronta di un punzone rigo su un materiale non omogeneo, *Ingegniaria Meccanica*, **14**, No. 9, pp. 37–50.
[9] G. Favretti (1965) Dipendenza fra durezza e profondita di ci mentazione—Applicazione della teoria della plasticità dello studio teorico del problema, *Ingegniera Meccanica*, **15**, No. 6.
[10] G. Favretti (1966) Indentation of a rigid punch on a plastically non-homogeneous material, *Meccanica*, **1**, No. 3/4, pp. 83–94.
[11] P. Florentin and Y. Gabriel (1974) Force portante d'une fondation sur sol verticalement non-homogène, *Trav. fin d'études E.N.P.C.*, Lab. Mécanique des Solides, Ec. Polytech. Paris, june 1974.
[12] H. Geiringer (1953) Some recent results in the theory of an ideal plastic body, *Advances in Applied Mechanics*, Academic Press, New York.
[13] A. P. Green (1954) On the use of hodographs in problems of plane plastic strain, *J. Mech. Phys. Solids*, **2** No. 2, pp. 73–80.
[14] Bent Hansen. A theory of Plasticity for ideal frictionless material. Thesis.
[15] R. Hill (1950) *The Mathematical theory of Plasticity*, Clarendon Press, Oxford.
[16] R. G. James and P. L. Bransby (1971) A velocity field for some passive earth pressure problems. *Géotechnique*, **21**, No. 1, pp. 61–83.
[17] A. W. Jenike and R. T. Shield (1969) On the plastic flow of Coulomb solids beyond original failure, *Jl. Appl. Mech., Trans. A.S.M.E.*, **27**, pp. 599–602.
[18] A. I. Kuznetzov (1958) The problem of torsion and plane strain of non-homogeneous plastic bodies. *Arch. Mech. Stos.*, **4**, pp. 447–462.

[19] E. H. Lee (1950) On stress discontinuities in plane plastic flow. *Proc. 3rd Symp. Appl. Math.*, McGraw Hill ed., pp. 213–228.
[20] J. Najar, J. Rychlewski and G. S. Shapiro (1966) On the problems of the elastic-plastic state of an infinite wedge. *Bull. Ac. Pol. Sc.*, **14**, No. 9, pp. 515–522.
[21] J. P. Obin (1972) Force portante en déformation plane d'un sol verticalement non-homogène. Thesis Univ. Grenoble.
[22] W. Olszak and J. Rychlewski (1962) Geometrical properties of stress fields in plastically non-homogeneous bodies under conditions of plane strain. *Pr9c. Int. Symp. 2nd ord. In Elasticity, Plasticity and Fluid Dynamics*, Haifa.
[23] W. Olszak, J. Rychlewski and W. Urbanowski (1962) Plasticity under non homogeneous conditions. *Advances in Applied Mechanics*, Academic Press, New York, pp. 132–214.
[24] J. Ostrowska (1968) Initial plastic flow of semi-space with a strong layer non homogeneity. *Arch. Mech. Stos.*, **20**, No. 6, pp. 651–668.
[25] W. Prager (1955) The sign of plastic power in the graphical treatment of problems of plane plastic-flow. *Quart. Appl. Math.*, **13**, No. 3, pp. 333–335.
[26] J. Rychlewski (1966) Plane plastic flow for jump non-homogeneity. *Int. Jl. Non-Linear Mech.*, **1**, pp. 57–78.
[27] J. Salençon, M. Barbier and M. Beaubat (1973) Force portante d'une fondation sur sol non-homogène. *Proc. 8th Int. Conf. Soil Mech. & Found. Eng., Moscow*, **1**.3, pp. 219–224.
[28] J. Salençon (1974) Bearing capacity of a footing on a $\phi = 0$ soil with linearly varying shear strength. *Géotechnique*, **24**, No. 3, pp. 433–446.
[29] A. A. Serrano (1972) El metodo de los campos associados. *Proc. 5th Eur. Conf. Soil Mech.*, pp. 77–84.
[30] R. T. Shield (1953) Mixed boundary value problems in soil mechanics. *Quart. Appl. Math.*, **11**, pp. 61–75.
[31] W. Szczepinski (1971) Some slip-line solutions for earthmoving processes. *Archives of Mechanics*, **23**, No. 6, pp. 885–896.
[32] W. Szczepinski and H. Winek (1971) On some problems of large flow of soils. Symp. Franco-Polonais, Problèmes de la Rhéologie, Warsaw, 1971, pp. 353–365.
[33] A. Winzer and G. F. Carrier (1948) The interaction of discontinuities surfaces in plastic fields of stress. *Jl. Appl. Mech. Trans. ASME*, **15**, pp. 261–264.
[34] A. Winzer and G. F. Carrier (1949) Discontinuities of stress in plane plastic flow. *Jl. Appl. Mech., Trans. ASME*, **16**, pp. 346–348.

B. PROBLEMS OF UNCONTAINED PLASTIC FLOW WITH AXIAL SYMMETRY FOR MATERIALS WITH A YIELD CRITERION OF THE 'INTRINSIC CURVE' TYPE

1. General

This appendix indicates the type of plasticity problems occurring with axial symmetry for materials obeying a yield criterion of the 'intrinsic curve' type, under the hypothesis of Haar–Karman currently adopted in soil mechanics.

The theory already explained in Chapter IV and in the preceding appendix will not be repeated but some problems regarding discontinuity lines, that have been ignored so far, will be studied.

2. The Stress Problem

2.1 Statement of the Problem, Haar–Karman hypothesis

With r, ω, z as the cylindrical coordinates, consideration will be given to problems of uncontained plastic flow for which the stress distribution has the following properties.

(1). There is axial symmetry around Oz: σ is independent of ω.

(2). $\sigma_{r\omega} = \sigma_{z\omega} = 0$: $\sigma_{\omega\omega}$ is a principal stress.

The stress-field σ must satisfy the conditions of equilibrium, which in particular require that

(1). There is no tangential component of the body force: $F_\omega = 0$;
(2). The boundary conditions are compatible with $\sigma_{r\omega} = \sigma_{z\omega} = 0$ ($T_\omega = 0$ on the surfaces perpendicular to O_z and to O_r);
(3). The body forces and the boundary conditions are axially symmetric about Oz: F_r and F_z are independent of ω, and data T_r and T_z on a plane perpendicular to Oz are independent of ω.

The stress-field must therefore satisfy both equations of equilibrium, i.e.

$$\left. \begin{array}{l} \dfrac{\partial \sigma_r}{\partial r} + \dfrac{\partial \sigma_{rz}}{\partial z} + \dfrac{\sigma_r - \sigma_\omega}{r} + \rho F_r = 0 \\[6pt] \dfrac{\partial \sigma_{rz}}{\partial r} + \dfrac{\partial \sigma_{zz}}{\partial z} + \dfrac{\sigma_{rz}}{r} + \rho F_z = 0 \end{array} \right\} \quad (2)$$

The isotropic material is assumed to have a yield criterion of the intrinsic curve type. It is not necessarily homogeneous, but it is necessary, so that (1) is true, that the yield criterion is independent of ω.

With the usual notation

$$\sigma_1 \geqslant \sigma_2 \geqslant \sigma_3 \quad (3)$$

$$p = -(\sigma_1 + \sigma_3)/2 \quad (4)$$

$$R = (\sigma_1 - \sigma_3)/2 \quad (5)$$

the criterion assumes the form

$$R = R(p, r, z) \quad (6)$$

According to condition (1), σ_ω is a principal stress. The Haar–Karman hypothesis (6), which states that the *flow regime in a plastic zone occurs along an edge (or corner), σ_ω being equal to one of the principal stresses in the meridian plane*, gives

$$\sigma_\omega = \sigma_2$$

and

$$\left. (\sigma_2 = \sigma_1 \quad \text{or} \quad \sigma_2 = \sigma_3) \right\} \quad (7)$$

This hypothesis is discussed further in [3, 8, 12].

For $\theta = (Or, \sigma_1)$,

$$\left.\begin{array}{l}\sigma_r = -p + R\cos 2\theta \\ \sigma_z = -p - R\cos 2\theta \\ \sigma_{rz} = R\sin 2\theta \\ \sigma_\omega = -p - \varepsilon R\end{array}\right\} \quad (8)$$

where

$$\begin{array}{ll}\varepsilon = -1 & \text{corresponds to} \quad \sigma_\omega = \sigma_1, \\ \varepsilon = +1 & \sigma_\omega = \sigma_3\end{array} \quad (9)$$

(See [2], which has different sign conventions.)

In the plastic zones

$$R = R(p, r, z) \quad (6)$$

and

$$\partial R/\partial p = \sin\phi(p, r, z) \quad (10)$$

Hence, for the stress field in the plastic zone, the system of partial differential equations obtained by applying (6, 8, 10) to (2) is

$$-\frac{\partial p}{\partial r}(1 - \sin\phi\cos 2\theta) - 2R\sin 2\theta\frac{\partial \theta}{\partial r} + \sin\phi\sin 2\theta\frac{\partial p}{\partial z} + 2R\cos 2\theta\frac{\partial \theta}{\partial z}$$
$$+ \frac{\partial R}{\partial r}\cos 2\theta + \frac{\partial R}{\partial x}\sin 2\theta + R\frac{\cos 2\theta + \varepsilon}{r} + \rho F_r = 0 \quad (11)$$

$$\frac{\partial p}{\partial r}\sin\phi\sin 2\theta + 2R\cos 2\theta\frac{\partial \theta}{\partial r} - \frac{\partial p}{\partial z}(1 + \sin\phi\cos 2\theta) + 2R\sin 2\theta\frac{\partial \theta}{\partial z}$$
$$+ \frac{\partial R}{\partial r}\sin 2\theta - \frac{\partial R}{\partial z}\cos 2\theta + R\frac{\sin 2\theta}{r} + \rho F_z = 0 \quad (12)$$

The system (11, 12) is identical to that obtained in the Appendix for the plane problems, though two supplementary terms have appeared, which are due to axial symmetry and are taken into account as body forces.

The system (11, 12) is hyperbolic and quasi-linear. At each point of the meridian plane, there are two characteristic directions:

$$dz/dr = \tan[\theta \mp (\pi/4 + \phi/2)] \quad (\alpha, \beta \text{ lines}) \quad (13)$$

The relations along the characteristics are

$$dp + \frac{2R}{\cos\phi}d\theta - \left\{\rho F^\alpha + \frac{\partial R}{\partial x_\beta} - \frac{R}{r\cos\phi}\left[\sin\left(\theta - \left(\frac{\pi}{4} + \frac{\phi}{2}\right)\right)\right.\right.$$
$$\left.\left. - \varepsilon\sin\left(\theta + \left(\frac{\pi}{4} + \frac{\phi}{2}\right)\right)\right]\right\}dx^\alpha = 0 \quad (14)$$

$$dp - \frac{2R}{\cos\phi}d\theta - \left\{\rho F^\beta - \frac{\partial R}{\partial x_\alpha} + \frac{R}{r\cos\phi}\left[\sin\left(\theta + \left(\frac{\pi}{4} + \frac{\phi}{2}\right)\right)\right.\right.$$
$$\left.\left. - \varepsilon\sin\left(\theta - \left(\frac{\pi}{4} + \frac{\phi}{2}\right)\right)\right]\right\}dx^\beta = 0 \quad (15)$$

3. The Velocity Problem

The components of the velocity are denoted by u_r, u_ω, u_z. The normal relationships of radial symmetry are

$$v_{rr} = \frac{\partial u_r}{\partial r}, \quad v_{\omega\omega} = \frac{1}{r}\frac{\partial u_\omega}{\partial \omega} + \frac{u_r}{r}, \quad v_{zz} = \frac{\partial u_z}{\partial z}$$

$$2v_{\omega z} = \frac{\partial u_\omega}{\partial z} + \frac{1}{r}\frac{\partial u_z}{\partial \omega}, \quad 2v_{rz} = \frac{\partial u_z}{\partial r} + \frac{\partial u_r}{\partial z}, \quad (16)$$

$$2v_{r\omega} = \frac{1}{r}\frac{\partial u_r}{\partial \omega} + \frac{\partial u_\omega}{\partial r} - \frac{u_\omega}{r}$$

The stress problem having been solved in the plastic zones, the velocity problem will be solved in the same region.

3.1 Case of a standard material

The material is assumed to obey the principle of maximum plastic work, and v, which has the same principal directions as σ, has, as principal values,

$$\left.\begin{array}{l} v_1 = \left(\lambda + \frac{1+\varepsilon}{2}\mu\right)(1 + \sin\phi) \\[6pt] v_3 = -\left(\lambda + \frac{1-\varepsilon}{2}\mu\right)(1 - \sin\phi) \\[6pt] v_{\omega\omega} = -\varepsilon\mu(1 - \varepsilon\sin\phi) \end{array}\right\} \quad (17)$$

$$\lambda \geqslant 0, \mu \geqslant 0, \varepsilon = \pm 1 \quad (18)$$

3.1.1. Calculation of u_ω

With $v_{\omega\omega}$ being a principal value,

$$v_{r\omega} = v_{z\omega} = 0 \quad (19)$$

Assuming that the velocity field is also axially symmetric (on condition that this assumption be compatible with the boundary conditions), (16) and (19) yield

$$\frac{\partial u_\omega}{\partial z} = 0 \quad (20)$$

$$\frac{\partial u_\omega}{\partial r} - \frac{u_\omega}{r} = 0 \qquad (21)$$

whence, u_ω, with the form $u_\omega = \alpha r$, representing a rigid-body rotation about Oz.

$$(22)$$

3.1.2. In the plane (r, z)

For the unit basis tangent at each point to the lines α and β, equation (17) yields

$$v_{\alpha\alpha} = \varepsilon \frac{\mu}{2} \cos^2 \phi \qquad (23)$$

$$v_{\beta\beta} = \varepsilon \frac{\mu}{2} \cos^2 \phi \qquad (24)$$

$$v_{\alpha\beta} = \left(\lambda + \frac{\mu}{2}\right) \cos^2 \phi \qquad (25)$$

Taking μ from (17) and the expression for $v_{\omega\omega}$ from (16) into account gives the results

$$v_{\alpha\alpha} = -\frac{u_r}{2r}(1 + \varepsilon \sin \phi) \qquad (26)$$

$$v_{\beta\beta} = -\frac{u_r}{2r}(1 + \varepsilon \sin \phi) \qquad (27)$$

which show that the α and β lines are velocity characteristics. Equation (26) is a differential relation valid along the α (resp. 27, β) lines. These relations can also be written in terms of u_r and u_z from the classical formulae expressing $v_{\alpha\alpha}$ and $v_{\beta\beta}$ as functions of v_{rr}, v_{rz}, v_{zz} and from (16). Equations (26) and (27) are the relations along the velocity characteristics.

3.1.3. Conditions of positivity

As indicated in (18), λ and μ must be non-negative. Therefore, equation (25) implies that,

$$v_{\alpha\beta} \geqslant 0,$$

but this condition alone is not sufficient. From equations (17) and (22) is obtained,

$$\mu \geqslant 0 \Leftrightarrow \varepsilon \frac{u_r}{r} \leqslant 0 \qquad (28)$$

Then

$$\lambda \geqslant 0 \Leftrightarrow \left(\lambda + \frac{\mu}{2}\right) \cos^2 \phi \geqslant \frac{\mu}{2} \cos^2 \phi$$

or, from (25),

$$v_{\alpha\beta} \geq \frac{\mu}{2} \cos^2 \phi$$

and also

$$2v_{\alpha\beta} \geq - \varepsilon \frac{u_r}{r}(1 + \varepsilon \sin \phi) \qquad (29)$$

Equations (28) and (29) are the conditions of positivity.

3.2 The case of non-standard materials

In the case of non-standard materials those considered here have a plastic potential g of the intrinsic curve type (as in the Appendix on plane problems), which is assumed to be independent of ω. The angle of dilatancy v is a function of the stress state and of the point

$$0 \leq v(p, r, z) \leq \phi(p, r, z)$$

Then the strain-rate tensor has the same principal directions as $\boldsymbol{\sigma}$ and its principal values are

$$\left. \begin{array}{l} v_1 = \left(\lambda + \dfrac{1+\varepsilon}{2}\mu\right)(1 + \sin v) \\[6pt] v_3 = -\left(\lambda + \dfrac{1-\varepsilon}{2}\mu\right)(1 - \sin v) \\[6pt] v_{\omega\omega} = -\varepsilon\mu(1 - \varepsilon \sin v) \\[6pt] \lambda \geq 0, \mu \geq 0, \varepsilon = \pm 1 \end{array} \right\} \qquad (30)$$

Hence, following the hypothesis that the velocity field is axially symmetric

$$u_\omega = \alpha_r$$

The problem in the meridian plane is linear hyperbolic and at each point the characteristic directions are γ and δ:

$$\left. \begin{array}{l} (0_r, \gamma) \\ (0_r, \delta) \end{array} \right\} = \theta \mp \left(\frac{\pi}{4} + \frac{v}{2}\right) \qquad (31)$$

The relations along the velocity characteristics are given by (26, 27) in which (α, β, ϕ) are replaced by (γ, δ, v). The same argument applies for the conditions of positivity (28, 29).

4. Weak Solutions

4.1 Weak solutions for stresses

The properties of the weak solutions for stresses are analogous to those for plane problems. The discontinuity lines[1] are not the characteristics. The discontinuity condition expresses the continuity of the stress. $\sigma_{\omega\omega}$ is discontinuous.

4.2 Weak solutions for velocities

The lines of discontinuity of the velocity are the velocity characteristics, α, β in the case of a standard material, and γ, δ in the case of the non-standard material of Section 3.2. Expressed by means of equations (26, 27, 28, 29), this statement implies that

(1). For crossing of an α (resp. β, or γ resp. δ) line v_β is continuous (resp. α or δ resp. γ) and the discontinuity concerns only v_α (26, 27);

(2). Crossing an α (resp. β or γ resp. δ) line, $[v_\alpha] \geq 0$ (resp. β or γ resp. δ) (29), and $[v_\alpha]$ must be such that (28) is verified on both sides;

(3). The equation of propagation is also slightly modified with respect to the plane case, and becomes

$$D_\alpha[v_\alpha] = -\frac{1}{2r}[v_\alpha]\cos\left[\theta - \left(\frac{\pi}{4} + \frac{\phi}{2}\right)\right](1 + \varepsilon\sin\phi) \qquad (32)$$

since

$$[u_r] = [u_\alpha]\cos\left[\theta - \left(\frac{\pi}{4} + \frac{\phi}{2}\right)\right]$$

with an analogous equation along a β line (resp. γ and δ).

It follows (see [7, 12]) that a finite velocity discontinuity on the axis Oz is not propagated.

References

[1] J. F. Adie and J. M. Alexander (1967) A graphical method of obtaining hodographs for upper bound solutions to axisymmetric problems, *Int. Jl. Mech. Sc.*, **9**, No. 6, p. 349.
[2] B. G. Berezancew (1952) *Problème de l'Équilibre Limite d'un Milieu Pulvérulent en Symétrie Axiale*, Moscow.
[3] A. D. Cox, G. Eason, and H. G. Hopkins (1961) Axially symmetric plastic deformations in soils, *Phil. Trans. Roy. Soc. London, A*, **1036**, 254, pp. 1–45.
[4] M. Croc, G. Michel, and A. Pecker (1972) Quelques problèmes de non-homogénéité en symétrie axiale, *Trav. fin d'études E.N.P.C.*, Lab. Mécanique des Solides, Ec. Polytechn. Paris.
[5] G. Eason and R. T. Shield (1960) The plastic indentation of a semi-infinite solid by a perfectly rough circular punch. *Z.A.M.P.*, **11**, pp. 33–43.

---

[1] These are the traces in the meridian plane of the surfaces of discontinuity of the axially symmetric stress-field.

[6] A. Haar and Th. Karman (1909) Zur theorie der Spannungszustände in plastischen und sandartigen Medien, *Nachr. Ges. Wiss. Göttingen, Math. Phys. Kl.*, pp. 204–218.

[7] D. D. Ivlev and R. I. Nepershin (1973) Impression of smooth indenter into a rigid plastic half-plane, *Izv. AN. SSSR, Mekhanika Tverdogo Tela*, **8**, No. 4, pp. 159–163, *Engl. Transl. Mechanics of Solids*, pp. 144–149.

[8] J. Mandel and F. Parsy (1961) Quelques problèmes tridimensionnels de la théorie du corps parfaitement plastique, *Sém. Plasticité*, Ec. Polytechnique, P.S.T., No. 116, pp. 105–127.

[9] Z. Mroz (1967) Graphical solution of axially symmetric problems of plastic flow, *Z.A.M.P.*, **18**, pp. 219–236.

[10] R. Negre (1968) Contribution à l'étude de l'équilibre limite des sols et des matériaux pulvérulents et cohérents, Thesis Dr. Sc., Grenoble.

[11] J. Salençon, M. Croc, G. Michel and A. Pecker (1973) Force portante d'une fondation de revolution sur un bicouche, *C.R. Ac. Sc., Paris, série A*, **276**, pp. 1569–1572.

[12] R. T. Shield (1955) On une plastic flow of metals under conditions of axial symmetry, *Proc. Roy. Soc.*, **233**, A, 1183, pp. 267–287.

[13] R. T. Shield (1955) Plastic flow in a converging conical channel, *Jl. Mech. Phys. Solids*, **3**, pp. 246–258.

[14] R. Sibille (1944) Calcul approché des solutions de Prandtl dans les cas de révolution, *C.R. Ac. Sc. Paris*, t. 258, gr. 2, pp. 2017–2019.

[15] A. J. M. Spencer (1964) The approximate solution of certain problems of axially symmetric plastic flow, *Jl. Mech. Phys. Solids*, **12**, No. 4, pp. 231–243.

[16] W. Szczepinski (1967) Wstep do analisy procesow obrohki, *Inst. Podst. Probl. Techn., Pol. Ac. Sc.*

[17] W. Szczepinski, L. Dietrich, E. Drescher, and J. Miatowski (1966) Plastic flow of axially symmetric notched bars pulled in tension, *Int. Jl. Solids and Structures* **2**, pp. 543–554.

CHAPTER V

The Theory of Limit Analysis (For Applications to Soil Mechanics)

1. Presentation

This chapter deals with the theory of limit analysis and its applications to Soil Mechanics. No mention will be made (except for didactic purposes) of the problems related to the application of this theory to the design of structures, i.e. the so-called limit design of structures. For this purpose reference may be made, for instance, to [7, 8, 40].

The chosen exposition is classical and simplified (similar to that of [37]). A slightly more axiomatic and detailed presentation will be found in the Appendix, with the possible extensions of the theory to the case of non-standard materials [44, 45].

The concept of limit loading of a system was introduced in Chapters II and III, and here only the essentials of the previous argument will be repeated.

1.1 Definition of the limit loadings

By considering a system made of an elastic–perfectly plastic material, subject to a loading process depending on n parameters Q_i ($\mathbf{Q}$ = loading vector), the so-called *initial elastic limit load* Q^0 was defined, corresponding to the appearance of plastic deformations, for any loading path starting from the neutral state of stress. The set of all these loadings $\mathbf{Q}^0$ is the *initial elastic boundary of the system*. For the loading path going beyond $\mathbf{Q}^0$, the *limit loading* is defined

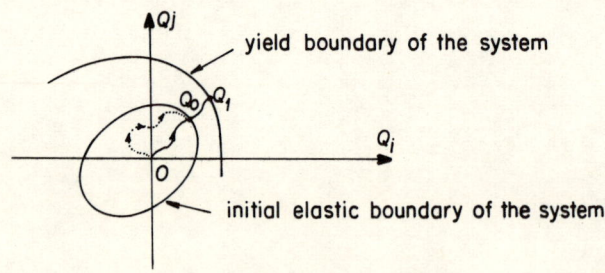

Figure V.1

as the loading which generates uncontained plastic flow for the *assumption of negligible geometry changes*. The set of limit loadings has been called the *yield boundary of the system* (Figure V.1) for reasons to be explained in this chapter.

It was shown in Chapter III that, since the geometry changes are assumed negligible up to the appearance of the uncontained plastic flow, the limit loading on a given loading path can be determined by considering the system as constituted by a rigid-plastic material (defined by passing to the limit) provided that this system follows the same loading path as the elasto-plastic system considered initially. Hence, the possibility arises of defining the limit loadings on the rigid-plastic system (obtained in each case by passing to the limit) as the loadings required for the appearance of uncontained plastic flow in the elasto-plastic system.

Finally, introduction of the rigid-plastic system obtained from the initial system by eliminating the elastic deformation of the material (without passing it to the limit) results in the limit loadings defined above being loadings for which there is a non-zero deformation of the associated rigid-plastic system.

In this chapter a study will be made first of the loadings for which there is a non-zero deformation of the associated rigid-plastic system. For convenience, these will be called limit loadings without specifying 'of the associated rigid-plastic system.' This omission will appear justified *a posteriori* in the case of systems satisfying the principle of maximum plastic work at any point.

1.2. Determination of the limit loadings. Variational approach

The determination of the limit load (in the case of a loading with one parameter) or of the yield boundary of the system (in the case of a loading with several parameters) requires the construction of complete solutions of the problem of uncontained plastic flow for the rigid–perfectly plastic system. This is often difficult, and hence a variational approach is used. Two theorems will be demonstrated, which make it possible to obtain under- and over-approximations.

2. Admissible Fields. Dissipation

The system is assumed to undergo a loading depending on n parameters Q_i, the boundary conditions being both dynamic (body forces and stresses) and kinematic (relating to the velocities) (see chapter III, Section 5.2).[1]

2.1 Plastically admissible stress tensor

At a point M of the system where the loading function is f, a stress tensor $\boldsymbol{\sigma}$ is said to be plastically admissible (P.A.) if

$$f(\boldsymbol{\sigma}) \leqslant 0 \tag{1}$$

---

[1] In this formulation, there are no constant data except those equal to zero. If some data are non-zero constant, they will be dealt with as though they were variable, and will be given their prescribed values at the end of the solution.

2.2 Plastically admissible strain rate tensor

At a point M of the system where the loading function is f, and where the plastic flow rule assumes the form (L)—which is not necessarily that of the standard material—a strain rate tensor $\mathbf{v}$ is said to be plastically admissible (P.A.) if the flow rule (L) may be solved for this tensor; i.e. if

$$\exists \boldsymbol{\sigma}, \; f(\boldsymbol{\sigma}) = 0 \quad \text{and} \quad \boldsymbol{\sigma} \xrightarrow{(L)} \mathbf{v} \qquad (2)$$

REMARK. This inversion is not always possible, and when possible, is not always unique. The example of a standard Mises material is a good illustration of this result. For the inversion to be possible, the given tensor $\mathbf{v}$ must satisfy $\mathrm{tr}(\mathbf{v}) = 0$, as the plastic deformation occurs without change in volume. If $\mathrm{tr}(\mathbf{v}) = 0$ the inversion is indeterminate and only the deviator s of $\boldsymbol{\sigma}$ is determined. In the case of the Tresca criterion, the degree of indeterminacy can be yet greater.

2.3 Dissipation

A stress tensor $\boldsymbol{\sigma}$ is associated with a plastically admissible strain rate tensor $\mathbf{v}$ by relation (2) in the expression

$$\sigma_{ij} v_{ij} = \boldsymbol{\sigma}\mathbf{v} \qquad (3)$$

Under the hypothesis of the principle of maximum plastic work (equivalent to f being convex and the material standard), the expression (3) depends only on $\mathbf{v}$ and is independent of the tensor $\boldsymbol{\sigma}$ associated with $\mathbf{v}$. Therefore,

$$\boldsymbol{\sigma}\mathbf{v} = \pi(\mathbf{v}) \qquad (4)$$

which is the dissipation for this strain rate tensor.

Proof:

Let $\boldsymbol{\sigma}^1$ and $\boldsymbol{\sigma}^2$ be two stress tensors associated with $\mathbf{v}$ by (2), in which f is convex and (L) is the flow rule for the standard material.

Then

$$\left. \begin{array}{llll} \text{for} & \boldsymbol{\sigma}^1 : f(\boldsymbol{\sigma}^1) = 0 & \mathbf{v} \in \lambda \partial f(\boldsymbol{\sigma}^1) & \lambda > 0 \\ \text{for} & \boldsymbol{\sigma}^2 : f(\boldsymbol{\sigma}^2) = 0 & \mathbf{v} \in \lambda \partial f(\boldsymbol{\sigma}^2) & \lambda > 0 \end{array} \right\} \qquad (5)$$

whence,

$$\mathbf{v}(\boldsymbol{\sigma}^1 - \boldsymbol{\sigma}^2) \geqslant 0 \quad \text{and} \quad \mathbf{v}(\boldsymbol{\sigma}^2 - \boldsymbol{\sigma}^1) \geqslant 0$$

It follows that

$$\boldsymbol{\sigma}^1 \mathbf{v} = \boldsymbol{\sigma}^2 \mathbf{v} = \pi(\mathbf{v})$$

In the case of a Mises material,

$$\pi(\mathbf{v}) = k\sqrt{2 v_{ij} v_{ij}} \qquad (6)$$

In the case of a Tresca material, which has the flow rule

$$\left.\begin{array}{l} \sigma_1 - \sigma_3 = 2k \\ \sigma_1 > \sigma_2 > \sigma_3 \\ v_1 = \lambda \\ v_2 = 0 \quad \lambda \geq 0 \\ v_3 = -\lambda \end{array}\right\} \quad (7a) \qquad \left.\begin{array}{l} \sigma_1 - \sigma_3 = 2k \\ \sigma_1 = \sigma_2 > \sigma_3 \\ v_1 = \lambda \quad \lambda \geq 0 \\ v_2 = \mu \quad \mu \geq 0 \\ v_3 = -\lambda - \mu \end{array}\right\} \quad (7b)$$

Side face regime \qquad\qquad\qquad Edge (or corner) regime

$$\pi(\mathbf{v}) = k\{|v_1| + |v_2| + |v_3|\} \tag{8}$$

2.4 Admissible stress-field

A stress-field is said to be admissible if it is statically and plastically admissible (S.P.A.).

S.A.: the stress-field must satisfy the equations of equilibrium and the boundary conditions for the stresses (dynamic conditions of the problem).

P.A.: at any point of the system, the stress tensor must be plastically admissible.

2.5 Admissible strain rate field

A strain rate field is said to be admissible if it is kinematically and plastically admissible (K.P.A.).

K.A.: the considered strain-rate field v must derive from a velocity field u^1 such that

$$\exists u, \; v_{ij} = \tfrac{1}{2}(u_{ij} + u_{j,i}) \tag{9}$$

and this velocity field must satisfy the kinematic boundary conditions of the problem.

P.A.: at each point of the system, the strain-rate tensor must be plastically admissible.

2.6 Dissipation in an admissible strain-rate field

Under the hypothesis of the principle of maximum plastic work, the dissipation in an admissible strain rate field is defined by the integral

$$\int_V \pi(\mathbf{v}) \, dV \tag{10}$$

which is an univocal function of the field v.

---

[1] From hereon, the notation used is that $\boldsymbol{\sigma}, \mathbf{v}$ denote tensors, whilst σ, v denote fields of these tensors.

3. Static Approach

3.1 Admissible loadings

A loading **Q**, which can be equilibrated by at least one admissible stress-field, is called an admissible loading.

The set

$$\{\mathbf{Q}(\sigma) | \sigma \text{ admissible}\} \tag{11}$$

will be denoted by K. $K \subset R^n$.

It should be noted, in particular, that any limit loading is admissible. (Moreover, for the loading paths of Section 1.2, of an elasto-plastic structure for the assumption of small deformations, all the loadings, from 0 to $\mathbf{Q}^1$ inclusively, are admissible loadings.)

3.2 Convexity of K

THEOREM: If f is convex, then K is convex.

The proof is trivial. Let $\mathbf{Q}_1, \mathbf{Q}_2$ be two loadings of K, and σ_1, σ_2 two admissible fields associated with them. Let $\mathbf{Q}$ be given by $\mathbf{Q} = \lambda \mathbf{Q}_1 + (1 - \lambda)\mathbf{Q}_2, \lambda \in [0, 1]$. From the definition of loading parameters (Chapter III, Section 5.2), the stress-field

$$\sigma = \lambda \sigma_1 + (1 - \lambda)\sigma_2$$

is S.A. associated with **Q**.

Also, from the convexity of f, the stress at any point is limited by

$$f(\boldsymbol{\sigma}) = f(\lambda \boldsymbol{\sigma}_1 + (1 - \lambda)\boldsymbol{\sigma}_2) \leqslant 0 \tag{12}$$

since

$$f(\boldsymbol{\sigma}_1) \leqslant 0$$

and

$$f(\boldsymbol{\sigma}_2) \leqslant 0.$$

3.3 Points of K at infinity

THEOREM: For materials having a convex loading surface 'open' only in the direction of isotropic pressures (see Chapter I, Section 2.7), and with zero body forces applied to the system, K has a point at infinity in only one direction, at most.

Proof:

From the definition of the loading parameters (Chapter III, Section 5.2) and the fact that $\sigma = 0$ is plastically admissible—the material being non-hardening—the load $\mathbf{Q} = 0 \in K$.

It is assumed that K does have points at infinity, and (L) denotes the direction of one of these points. Since $0 \in K$ and K is convex, the whole semi-axis $0L \in K$.

Let $\mathbf{Q}^*$ be prescribed $\in 0L$.

To say that $0L \in K$ up to infinity implies that, out of all the admissible stress-fields associated with $\mathbf{Q}^*$, there is (at least) one which remains admissible if it is multiplied by λ(positive), however great it may be. Thus, this field must consist of an isotropic pressure at any point of the solid.

If there are no body forces this pressure is constant, and $\mathbf{Q}^*$ is a loading which corresponds to a uniform isotropic pressure throughout the solid. If it exists[1] then it is a loading proportional to one parameter, and proves the foreseen result.[2]

In the case of ductile materials, for which the loading surface is open in the directions of the isotropic pressures and tensions, K can have a point at infinity in two opposite directions.

3.4 Limit loadings

By definition, a limit loading $\mathbf{Q}$ is a loading for which a solution exists to the problem of uncontained plastic flow of the associated rigid-plastic system. Thus, there exists an admissible stress-field σ associated with $\mathbf{Q}$, and also an admissible strain rate field v to which the strain rate vector of the system $\dot{q}$ corresponds. (Also, as $v \neq 0$, $\dot{\mathbf{q}} \neq 0$.) The fields σ and v are associated by the flow rule.

3.5 Theorem of maximum work

It is assumed that the material constituting the system obeys the principle of maximum plastic work.

THEOREM: Let $\mathbf{Q}$ be a limit loading and $\dot{\mathbf{q}}$ the corresponding strain rate of the system. Let $\mathbf{Q}^*$ be an admissible loading; we have:

$$(\mathbf{Q} - \mathbf{Q}^*) \cdot \dot{\mathbf{q}} \geq 0 \tag{13}$$

Proof:

As σ and σ^* are the admissible fields associated with $\mathbf{Q}$ and $\mathbf{Q}^*$ and v denotes an admissible field associated with $\dot{\mathbf{q}}$, by definition,

$$(\mathbf{Q} - \mathbf{Q}^*)\dot{\mathbf{q}} = \int_V (\sigma - \sigma^*)\mathbf{v}\, dV$$

which is positive, from the assumption concerning the principle of maximum plastic work.

---

[1] It may be that the loading process under study admits no uniform pressure field in the solid as being statically admissible.

[2] The result is true in the case of non-zero constant body forces. For body forces variable as functions of m parameters, the directions of the points at infinity, if they exist, are inside the pyramid built upon the corresponding positive semi-axes.

3.6 Consequences

THEOREM: Under the hypothesis of maximum plastic work, any limit loading belongs to the boundary of *K*.

Proof:

Let **Q** be a limit loading point and **q̇** the corresponding velocity vector. It is known that $\mathbf{Q} \in K$.

If it is assumed that **Q** is inside *K*, then $\exists_\eta > 0$, so that $B(\mathbf{Q}, \eta) \subset K$. [$B(Q, \eta)$ = sphere with a centre *Q* and a radius η.] There can be found in *B*, and hence in *K*, a point **Q*** such that $(\mathbf{Q} - \mathbf{Q}^*)\mathbf{\dot{q}} < 0$, which is obviously absurd. Therefore, *Q belongs to the boundary of K*.

Also, the theorem of maximum work indicates that at the point of limit loading **Q**, **q̇** *is orientated along an outward normal to the yield boundary*, which, according to Section 3.2, is a convex surface.

3.7 Converse of the preceding theorem

Under the hypothesis of maximum plastic work, let **Q** be a point on the boundary of *K* (a convex surface). As shown in Section 3.3, $0 \in K$. For a proportional loading process (i.e. a radius issuing from 0) defined by the direction **0Q**, there exists a limit loading point[1] situated on the boundary, according to the theorem of Section 3.6. This point must be **Q**, as a consequence of the convexity. Thus, any loading belonging to the boundary of *K* is a limit loading.

3.8 Conclusions. Gvozdev's theorem

Under the hypothesis of the principle of maximum plastic work, the boundary of the convex *K* is the yield boundary of the *associated rigid-plastic system* and it also is a *convex surface*.

This result is the so-called 'theorem of uniqueness of the limit loads' and implies, amongst other things, that a loading situated on the boundary of *K* is the limit loading of any loading path that arrives there.[2] Concerning the elasto-plastic and the rigid-plastic systems (the latter being defined by passing to the limit in each case), the preceding theorem gives the following results.

---

[1] Here an assumption is made which can be considered as a fact of experience. For the experiment of proportional loading of an elasto-plastic system (under the assumption of small deformations), account being taken that the limit loadings of this system are limit loadings for the associated rigid-plastic system, it is evident that either:

(1). It is possible to go on loading endlessly, which implies that the set *K* has a point at infinity in the loading direction, or

(2). It is not possible to go on loading endlessly, and it must be admitted that this always happens as a consequence of the appearance of uncontained plastic flow; i.e. there exists a limit loading.

[2] When dynamic data are fixed at their prescribed values, a yield boundary is obtained for the other loading parameters which is the intersection of the boundary of *K* with a $(n-p)$ hyperplane of R^n, where *p* is the number of fixed loading parameters. Hence the properties of this yield boundary are known.

(1). From the viewpoint of limit loadings, there is no longer any reason for distinguishing between the rigid-plastic systems, as they are all equivalent to the associated rigid-plastic system.

(2). The limit loadings are, as a consequence, independent of the elastic properties of the initial material.

(3). The limit loadings being independent of the loading paths, it is, therefore, unnecessary to specify the initial stress state at the beginning of a loading process.

The following theorem forms the basis of the under-approximation of the limit loading.

GVOZDEV'S THEOREM: *A loading which can be equilibrated by an allowable stress-field is beyond or on the yield boundary.*[1]

In the case of a proportional loading, $\lambda = 0$, being allowable, the statement becomes

'Any value of λ, such that an allowable stress field can be associated with it, is smaller than or equal to the limit loading'.

λ_1 *is the greatest value of λ that an allowable stress-field can be associated with.*

3.9 Static method

Taking Gvozdev's theorem as a basis, together with the convexity of K, gives *a static method* for the determination of the limit loads and the yield boundary.

The construction of allowable stress-fields permits the observation of allowable loadings; and the largest convex polygon with those loadings as summits is an *inner approximation of the yield boundary* (Figure V.2).

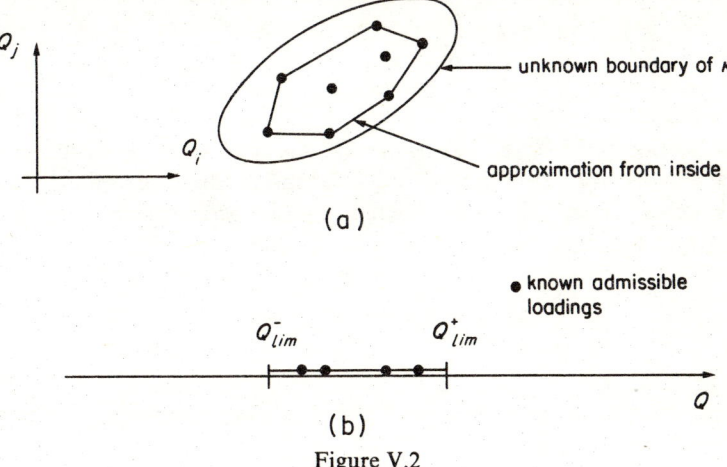

Figure V.2

---

[1] This theorem is more general than that given by Gvozdev [24] for the analysis of structures, taking as a basis the model due to Kazinczy [33] and Kist [34] (see [56]). The present statement, linked to the principle of maximum plastic work, seems to be due to Hill [26], in the form of a maximum principle.

4. Kinematic Approach

4.1 Theorem

Under the hypothesis of maximum plastic work, for a given permissible strain rate field v (having a corresponding strain rate of the system $\dot{q}$), and a permissible loading of the system $\mathbf{Q}^*$, *the power of $\mathbf{Q}^*$ in $\dot{q}$ cannot exceed the dissipation within the system of the strain-rate field v.*

Proof:

Let σ^* be an allowable stress field associated with $\mathbf{Q}^*$. Then, by definition,

$$\mathbf{Q}^*\dot{\mathbf{q}} = \int_V \sigma^* \mathbf{v} \, dV. \tag{14}$$

From the principle of maximum work, at each point of the system

$$\sigma^* \mathbf{v} \leqslant \sigma \mathbf{v} = \pi(\mathbf{v}) \tag{15}$$

where σ is a stress tensor resulting from the inversion of the flow rule for v. Hence, the result,

$$\mathbf{Q}^*\dot{\mathbf{q}} \leqslant \int_V \pi(\mathbf{v}) \, dV \tag{16}$$

4.2 Dual definition of K

For all the allowable strain rate fields, let K_1 denote a convex surface of $\{\mathbf{Q}\}$ space, the intersection of the half-spaces defined by

$$\mathbf{Q}\dot{\mathbf{q}}(v) \leqslant \int_V \pi(\mathbf{v}) \, dV \tag{17}$$

From Section 4.1 $K \subset K_1$. Consider $\mathbf{Q}^* \notin K$ and let $\mathbf{Q}$ be the point of K's boundary (convex) situated on $0\mathbf{Q}^*$. $\mathbf{Q}$ is a limit loading, according to Section 3.7, to which a permissible strain rate field v and a permissible strain rate of the system $\mathbf{q}(v) \neq 0$ correspond. According to the convexity of K

$$\mathbf{Q}^*\dot{\mathbf{q}}(v) > \mathbf{Q}\dot{\mathbf{q}}(v) = \int_V \pi(\mathbf{v}) \, dV$$

so that $\mathbf{Q}^* \notin K_1$.

It is possible to derive that $K = K_1$, and the yield boundary is the overall envelope of the planes given by

$$\mathbf{Q}\dot{\mathbf{q}}(v) = \int_V \pi(\mathbf{v}) \, dV \tag{18}$$

corresponding to all the allowable velocity fields.

If an allowable strain rate field for the system is known, then the equality

$$\mathbf{Q}\dot{\mathbf{q}}(v) = \int_V \pi(\mathbf{v})\,dV \tag{18}$$

determines a plane which is entirely exterior, or tangential to the yield boundary of the system.

For a plane which is a tangent to the boundary surface, the point of contact is a limit loading, corresponding to the strain rate $\dot{\mathbf{q}}$ which appears in (18).

In the case of a proportional loading, the value of λ obtained by balancing the work done by the exterior forces and the dissipation in any admissible velocity field is greater than or equal to the limit load:

$$\lambda = \frac{\int_V \pi(\mathbf{v})\,dV}{\mathbf{f}\cdot\mathbf{u}} \geq \lambda_1$$

Here, λ_1 is the smallest value of λ such that the work done by the exterior forces is equal to the dissipation in at least one allowable velocity field.

4.3 Kinematic approach

As in the static approach, the preceding results are used for the *kinematic method* of determining the limit loadings and the yield boundary. Allowable velocity fields are constructed and the convex polygon limiting the intersection of the half-spaces defined by (17) is an *exterior approximation of the yield boundary* (Figure V.3).

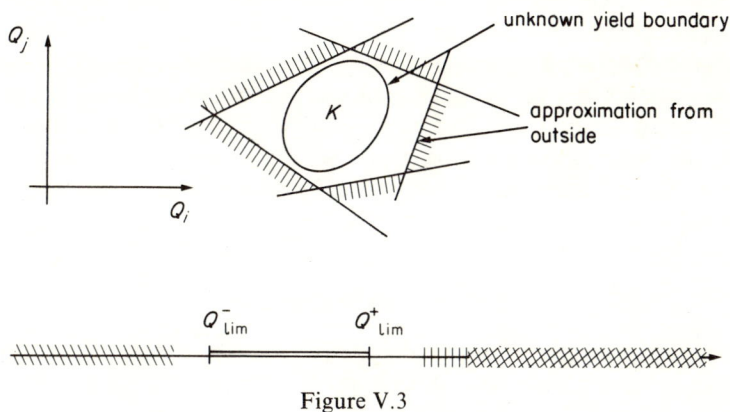

Figure V.3

5. Remarks on the Results of the Theory of Limit Analysis

5.1

From the logical viewpoint, the results of Section 3.6 prove that the yield boundary is either the whole or part of the K boundary, which is fundamental to the static method. Similarly, the results of Section 4.1 contain the basis of the kinematic method. However, it seems desirable to invoke the existence hypothesis in order to arrive at simpler and more 'natural' statements.

5.2

The results of the theory of limit loads, also termed limit analysis, can be expressed in the following form.

(1). If a loading is such that it is possible to determine an allowable stress-field which produces equilibrium, then the structure will resist this load.
(2). If an allowable deformation mechanism and a loading sufficient to cause collapse are known, then the structure will not resist this loading.

These results may appear intuitive, but as seen previously their proof requires that the principle of maximum plastic work be verified.[1] It should be noted that counterexamples of the theorems of Sections 3.8 and 4.2 have been given in [15, 49, 51].

5.3

In the case where the principle of maximum plastic work is not verified, whether because of the constitutive materials of the structure or because of the interface conditions between the different constitutive solids (e.g. Coulomb friction), the yield boundary can no longer be proved to be of $(n-1)$ dimensions. Stated otherwise, the 'theorem of uniqueness of the limit loads' is no longer applicable.

The independence, with respect to the initial stresses, the loading path and the elastic properties, of the limit loadings of the elasto-plastic system is no longer ensured. Also, it can no longer be stated that all the definitions of the rigid-plastic material are equivalent with respect to the limit loadings.

Non-standard systems have been studied extensively and Appendix A gives the results [5, 15, 31, 32, 42, 44–46, 50, 52, 55].

6. Minimum Principles

Using the principle of maximum plastic work, the classical minimum principles are given. They are valid for the case where the boundary conditions may be expressed in the classical form:

T given = $\mathbf{T}^d$ on S_T;
F given
u given = $\mathbf{u}^d$ on S_u

---

[1] The principle must be verified also at the interfaces of the solids that constitute the system.

6.1 Minimum principle for stresses (Hill)

Let σ be an allowable stress-field, to be used in the functional

$$\mathcal{H}(\sigma) = - \int_{S_u} \mathbf{T}(\sigma) \cdot \mathbf{u}^d \, dS_u \tag{19}$$

Then, *if a solution of the uncontained plastic flow problem exists for the given data, the accompanying stress-field renders the functional $\mathcal{H}$ a minimum among all the allowable stress-fields.*

6.2 Minimum principle for the strain rates (Markov).

With v being an allowable strain rate field, derived from the velocity field u, a functional is defined by

$$\mathcal{B}(v) = \int_V [\pi(\mathbf{v}) - \rho \mathbf{F} \cdot \mathbf{u}] \, dV - \int_{S_T} \mathbf{T}^d \cdot \mathbf{u} \, dS_T. \tag{20}$$

Then, *if a solution of the uncontained flow problem exists for these data, the accompanying strain rate field renders the functional B a minimum among all the allowable strain rate fields.*

The proof of these two principles is simple as they are supported by the theorem of virtual work and the principle of maximum plastic work. It can also be proved that for a solution of the uncontained flow problem

$$\mathcal{B}(v) = - \mathcal{H}(\sigma) \tag{21}$$

7. Limit Analysis in the Study of Plane Strain Problems of Uncontained Plastic flow

In the study of plane strain problems of uncontained plastic flow the exact significance of the solutions obtained was left aside. In particular, the fact that attention was paid only exceptionally to the undeformed zones is important.

It is now possible to deal with these points, assuming the principle of maximum plastic work to be valid, by means of the theory of limit analysis. Also, the uniqueness of the stress-fields proved in Chapter III (Section 6) makes it possible to derive further conclusions.

7.1 An example study

The example considered is the passive pressure of a wall on a wedge of Tresca material, as studied in Chapter IV. Two loading parameters are the components given by reducing the external forces acting on the rigid wall so as to act at a fixed point (e.g. the middle of the wall).

$Q_1 = M$ (moment)
$Q_2 = N$ (normal force—the tangential force is zero, as the wall is smooth)

A third parameter, Q_3, which is maintained constant, is given by the value of the surface load on OY (Figure V.4).

This problem has been partially solved to give allowable strain rate and stress-fields. The strain rate is known for the whole solid: above $ABCD$ $v \neq 0$, and under $ABCD$ $u = 0$, with a velocity discontinuity along $ABCD$. A stress-field σ (in limit equilibrium) is known above $ABCD$. At each point of the deformed zone σ and v are associated by the flow rule.

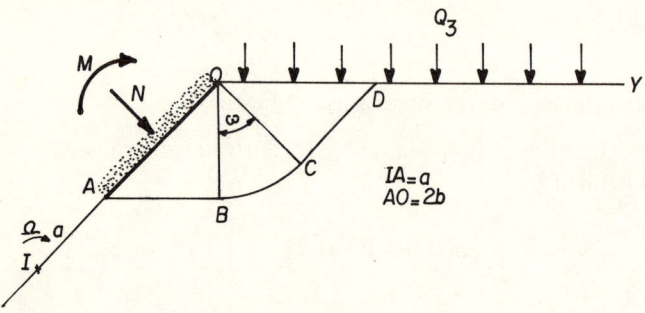

Figure V.4

The following comments may be made about such a solution.

(1). It is not complete as the determination of the stresses under $ABCD$ has not been dealt with.

(2). It is possible to use the kinematic method, as an allowable strain rate field has been provided for the system. The calculation of the dissipation integral is simplified by the knowledge of the allowable field σ associated with v in the deformed zone.

The work done by σ in v is equal to the dissipation $\pi(v)$.

By applying the theorem of virtual work[1] it is shown that the integral of the dissipation (including that along the velocity discontinuity line) is equal to the work done by the pressure Q_3 and the stresses under the rigid wall corresponding to σ.

By applying (18), the equation of the corresponding plane which is tangential or exterior to the yield boundary is obtained. It appears as an upper bound for the moment, with respect to I, of the forces applied to the plate.

(3). It is possible to show [48] that if the value of the surface load Q_3 is not too great, then an allowable extension of the stress field into the zone situated beneath $ABCD$ may be found.

Under these conditions, the static method can be applied, as an allowable stress-field σ is known. The corresponding loading is also allowable, and its representative point is not exterior to the yield boundary.

---

[1] This application is possible for the whole structure, although the stress-field in the undeformed zone remains unknown, as the equilibrium of the boundary forces is satisfied here.

By combining these results, it may be seen that

(1). An upper bound has been obtained for the moment, with respect to I, of the forces applied to $0A$:

$$M_I = M + N(a + b) \leqslant 2b[Q_3 + 2k(1 + \omega)](a + b) \tag{22}$$

(2). If Q_3 is not too great a limit loading has been obtained corresponding to the rotation of the plate around the point I:

$$\left.\begin{array}{l} M = 0 \\ N = 2b[Q_3 + 2k(1 + \omega)] \end{array}\right\} \tag{23}$$

For the case where $\omega = \pi/2$ and $Q_3 = 0$, the problem becomes the indentation of a half-plane without surface load. Three different extensions of the stress-field are known in this case [3, 57, 58]. The limit force of indentation becomes

$$F = 2b \times (\pi + 2)k \tag{24}$$

In the case where Q_3 is not too great, and where it is possible to extend the stress-field, a solution of the problem of uncontained plastic flow is obtained. Then the theorem of uniqueness for the stress-field, as given in Chapter III (Section 6), can be applied.

The stress-field in the deformed zone is shown to be unique for the given value of $\mathbf{Q}$, i.e. (23), or the given value of $\dot{\mathbf{q}}$, i.e.

$$\dot{q}_1 = \Omega$$
$$\dot{q}_2 = \Omega(a + b)$$
$$[(\dot{q}_3 = 2b\,\dot{q}_2 = 2b\Omega(a + b)].$$

Obviously, the results have more practical interest in the latter form, which implies that the stress distribution in $OABCD$, and in particular under OA, is the exact distribution. Thus, the exact distribution of stresses beneath OA is a uniform pressure equal to $[Q_3 + 2k(1 + \omega)]$.

7.2 General case

For plane strain problems, assuming the principle of maximum plastic work, the following general statements apply.

(1). If the proposed solution provides only a stress-field in limit equilibrium, in only a part of the system, then no conclusion can be reached concerning the result obtained.

(2). If the proposed solution provides a stress-field in limit equilibrium in a part D of the system, such that the static equilbrium of the boundary data for the remainder of the system is ensured, and an allowable velocity field is associated with the stress-field in D, leaving undeformed the exterior of D, then the solution is said to be *incomplete* [3]. This corresponds to the application of the kinematic method, as demonstrated in the example.

The result is an upper bound of the limit load for the case where there is only one loading parameter, or more generally, a hyperplane exterior to the yield boundary.

(3). If the solution provides a stress-field in limit equilibrium in a plastic zone, and an allowable extension of this stress-field outside this zone, then it is a *static* solution, making possible the application of the static method.

The result obtained is a lower bound of the limit load or, more generally, an allowable loading represented by a point internal to the yield boundary.

(4). If the proposed solution corresponds to both (2) and (3), the solution is *complete*, and provides the limit load, or more generally, the limit loading, and an associated deformation mechanism.

In fact, only complete solutions warrant the name solution. However, use of the above terminology, due to Bishop, has the advantage of being very explicit.

For complete solutions, the theorem of uniqueness of the stress-field is applicable, and demonstrates that the stress distribution in the deformed zones is the exact distribution for the given problem. This does not entail the uniqueness of the complete solutions, e.g. for the same incomplete solution several allowable extensions of the stress-field may be found to complete it.

The construction of the allowable extension of the stress-field of an incomplete solution is of no interest in itself. What is important is its feasibility. It is sufficient to have a theorem which proves that the extension is possible. At first, it might seem that such a theorem would be simple, since the stress-fields of the extension have only to verify the equilibrium equations and fit the stress boundary conditions, whilst not violating the yield criterion. However, a theorem does not exist, and it is usually necessary to actually construct an allowable extension. In fact, theorems are available which sometimes show that the extension is certainly not possible, e.g. Hill's theorem [28], and Bonneau's theorem (see Appendix B).

Following the work of Bishop [3] and Shield [58], research interest has in latter years been devoted to the static approach. A number of non-trivial static solutions and complete solutions have been demonstrated, though no general method for their construction exists [1, 9–11, 18, 25, 47, 48, 53, 54, 59].

Finally, although being obvious, it must be stressed that there is no uniqueness of incomplete solutions. These can give quite different results; and (2) indicates that the one that gives the best upper bound of the limit load (i.e. the lowest one) must be taken into consideration, but should not be thought to give the actual limit load. Also, it does not necessarily follow that the solution can be completed.

Incomplete solutions constitute most of the 'solutions' proposed for plane flow problems with a Tresca material.

7.3 Conclusions

Limit analysis acts as a guide in the research for 'solutions' of problems of uncontained (and particularly plane) flow for rigid-plastic materials. It is also the only method of choosing between several incomplete 'solutions'.

The mode of construction of the 'solutions', which may at first have appeared intuitive, and the distinction operated in Chapter III (Section 4) between a-type and b-type zones can now be better understood. Usually, the construction of an incomplete solution is planned by firstly forming an appropriate deformable (a-type) zone, and then ensuring the overall equilibrium of each b-type zone. Such a 'solution' is interpreted completely by the theory of limit analysis, as already stated. One may try then to complete the solution by an allowable extension of the stress-field into the b-type zones.

It should be noted that problems do exist where purely static or *kinematic* 'solutions' are developed. An example of those will be given in Section 8. This last type of solution, which has not been presented in Section 7.2, sometimes utilizes the theory of plane limit equilibrium and provides only an allowable mechanism of deformation, which may be interpreted by means of the theorems of Section 4.2.

8. Interpretation of the Solutions for the Case of a Coulomb Material

Soils obeying the Coulomb criterion do not satisfy the principle of maximum plastic work. The theory of limit analysis in its classical form is, therefore, not applicable.

The uniqueness theorem of the limit loadings is no longer applicable, as the set of limit loadings of the associated rigid-plastic system is no longer identical to the boundary of the set of allowable loadings.

Thus it may be possible for the same loading to be a limit loading for the elasto-plastic system along one loading path, but not along another. Similarly, it is possible that the same loading along a single loading path, may or may not be a limit loading for different elastic properties of the material. Finally, there is obviously no longer an equivalence between all the definitions of a rigid-plastic material.

With regard to the different types of solutions to the problems of plane deformation, and their significance, the analysis of Section 7 is no longer applicable. In most cases, the proposed solutions are analogous to those of Section 22 in Chapter IV, giving a stress-field in limit equilibrium in a part of the structure under investigation. They belong to type (1) of Section 7.2 and no conclusions can be made about the results obtained.

A position that may be adopted [38] for the interpretation of solutions in the case of a non-standard Coulomb material is to refer to the corresponding solutions to the same problems in the case of a standard Tresca material. For these, the interpretation can be supported by the theory of limit analysis. It is assumed that the interpretation for a Coulomb material is similar, without specifying the flow rule, as has been suggested previously in the construction of solutions.

The results obtained should be considered as approximate but, as often shown by experience, can be useful sources of information.

8.1 A Standard Coulomb material

The study of the case of a non-standard material with a convex yield criterion, presented in Appendix A of Chapter IV (also [55]), suggests ideas of great importance.

If f is the convex yield criterion of the material (not necessarily homogeneous), then the allowable loadings are defined as in Section 3.1, and constitute a convex set K_F. For any system with a criterion f, regardless of its elastic properties or flow rule, it is known that the loads which the system can bear and the limit loadings for all the loading paths necessarily lie within K_F. Only in the case of a standard system is it possible to state that the system can bear all the loads within K_F and that the boundary of K_F is the locus of the limit loadings.

For the determination of K_F and its boundary, which represent 'the maximum the system can do', all the available methods will be used: the static method founded on the convexity, and the kinematic method, assuming the material to be standard, which premise has no other significance than that of being a means of calculation. Following this reasoning, the solutions of plane problems will be interpreted assuming a standard material.

If a static extension of the stress-field is known, a static solution for the standard Coulomb material is obtained, and hence, an inner approximation to the boundary of K_F. If a velocity field can be associated with the stress field, an incomplete solution is found for the standard Coulomb material, and hence, an outer approximation to the boundary of K_F. Finally, a complete solution for the standard Coulomb material provides a point on the boundary of K_F and an associated tangent plane.

8.2 General remarks

Although not proved, it is thought that the method of interpretation proposed in Section 8.1 by reference to a standard Tresca material is in fact equivalent to that indicated previously, i.e. assuming that a Coulomb material is standard.

As indicated in Chapter IV, Section 23.4, the proof of the method of superposition considers static solutions for a Coulomb material, and is supported by the 'static theorem' and the static method. Therefore, it implicitly assumes that a Coulomb material is standard.

The allowable extensions of the stress fields used in the example have been given in [53, 58] for the case of a weightless material with cohesion and surface load, and in [10] for the case of a 'material with weight and without cohesion or surface load'.

It is possible to refer to a kinematic method for a non-standard material [55], since the construction of kinematic solutions for a standard material provides an outside approximation to all the possible limit loadings for a non-standard material. However, it would not be reasonable to speak of a static method for a non-standard material, especially relating to static solutions constructed for a standard Coulomb material with a criterion f.

8.3 Radenkovic's theorem

A theorem originated by Radenkovic [44, 45], and also proposed under various forms in [5, 31, 32, 42, 46, 52, 55] makes it possible to refer to static methods in the case of non-standard materials whose flow rules have particular properties [55]. This leads to an inner approximation of the zone of possible limit loadings.

The detailed proof of this theorem is given in Appendix A of the present chapter and in reference [55]. It supposes that the flow rule of the non-standard material derives from a plastic potential g which is convex, and has certain properties with respect to the criterion f. It is proved that the limit loadings for the non-standard material cannot lie within the set K_G of allowable loadings for the standard system constituted by a (fictitious) material with a criterion g. Thus, the boundary of K_G provides an internal boundary for the zone of possible limit loadings for the non-standard material, and hence, by means of the static method for a system made of the standard material with criterion g, it gives an internal approximation to all possible limit loadings for the non-standard material.

This theorem is applicable in particular to a Coulomb material which has a flow rule defined by an angle of dilation v, where $0 \leqslant v \leqslant \phi$ (see Chapter I). The potential g is then a Coulomb potential, with an angle v, which must be internal to the criterion f in the region where it is used.

Unfortunately, the results obtained by means of this theorem are two conservative to be of practical use, particularly in the case of $v = 0$.

8.4 The concept of a possible solution

When considering a Coulomb material with an angle of internal friction ϕ and a dilation angle v, it is sometimes possible to construct a static solution with which a velocity field is associated via the flow rule of a non-standard material. A complete solution for the non-standard material is then available. It is said in this case that a *possible solution* for the non-standard material is obtained, with the corresponding loading being a possible limit loading.

The example dealt with in Appendix A of Chapter IV (Section 6) is a complete solution for a standard Coulomb material f [13, 14, 53, 58]; whence the interpretation given in Section 8.2. However, if $v \neq \phi$, there does not exist any velocity field associated with the stress-field.[1] (The condition $\lambda \geqslant 0$ is not satisfied [13, 14].) An example of a possible solution for a non-standard material, for at least some values of the parameters, is to be found in [9].

[1] This does not prove that the corresponding loading is not a possible limit loading for $v \neq \phi$.

9. Other Applications of Limit Analysis in Soil Mechanics

The use of the theory of limit analysis in Soil Mechanics is not exclusively linked to problems of plane limit equilibrium such as those dealt with in Chapter IV. The theorems of both static and kinematic methods are variously applied in engineering practice.

9.1 Example

An an example, the determination of the maximum height of a vertical bank, consisting of a Tresca material with self-weight, is considered.

(a) Utilization of the static method

The static method is used as follows.

Any height h such that an allowable stress-field in the whole medium can be found is less than or equal to the limit height h_1. The allowable stress-field must satisfy the equilibrium equations,

$$\begin{cases} \dfrac{\partial \sigma_x}{\partial x} + \dfrac{\partial \tau_{xy}}{\partial y} + y = 0 \\ \dfrac{\partial \tau_{xy}}{\partial x} + \dfrac{\partial \sigma_y}{\partial y} = 0 \end{cases}$$

the 'plastically admissible' condition,

$$|\sigma_1 - \sigma_2| \leqslant 2k$$

and the stress boundary conditions,

$$\begin{cases} x = 0 \; y > 0: \sigma_x = \tau_{xy} = 0 \\ x = h \; y < 0: \sigma_x = \tau_{xy} = 0 \\ 0 < x < h \; y = 0: \sigma_y = \tau_{xy} = 0 \end{cases}$$

Conditions at infinity apply only to the velocities, which must be equal to zero.

The discontinuous stress-field proposed in [17] is represented in Figure V.5.

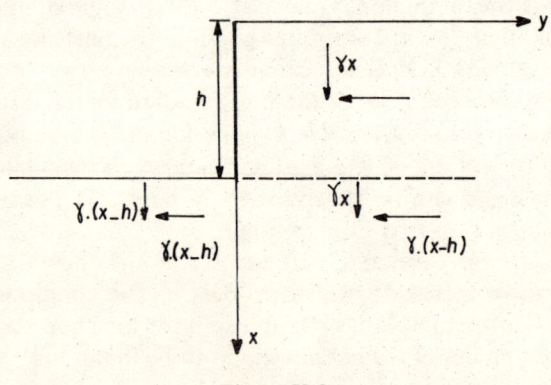

Figure V.5

It is defined by

$$y \geq 0 \quad 0 \leq x \leq h: \sigma_y = 0, \quad \tau_{xy} = 0, \sigma_x = -\gamma x$$
$$y \geq 0 \quad h \leq x \quad : \sigma_y = -\gamma(x-h), \tau_{xy} = 0, \sigma_x = -\gamma x$$
$$y \leq 0 \quad h \leq x \quad : \sigma_y = -\gamma(x-h), \tau_{xy} = 0, \sigma_x = -\gamma(x-h)$$

This field satisfies the equilibrium equations, the boundary conditions, and the continuity of the stress applied along $y = 0$, which is the line of discontinuity of the field. It is, therefore, statically admissible. It is also plastically admissible if

$$\gamma h \leq 2k$$

Therefore,

$$h = \frac{2k}{\gamma} \leq h_1$$

(b) Utilization of the kinematic method

The kinematic method is applied in the following way.

A height for which an allowable mode of deformation exists, where the power of the exterior forces (gravity) balances the dissipation, is greater than or equal to the limit height of the bank.

An allowable velocity field must be such that

(1). The deformation occurs without volume change,

(2). The boundary conditions for the velocity are satisfied, i.e. $u_x = u_y = 0$ at infinity.

For the mechanism shown in Figure V.6, a rigid body slides along an isolated, (necessarily) circular line with a centre O and a radius h.

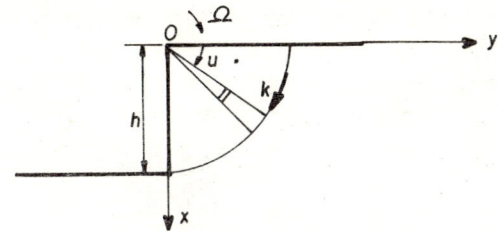

Figure V.6

This constitutes an allowable mode of deformation, and dissipation of the shear strength k occurs along the circle (friction), being equal to

$$k \times \Omega h \times \pi/2h$$

The work done by external (i.e. gravity) forces on the rotating block is equal to

$$\int_0^{\pi/2} du \int_0^h \gamma \Omega r \cdot (\cos u) \cdot dr = \gamma \Omega h^3/3$$

123

Hence, on balancing the two expressions,

$$h = \frac{3\pi}{2}\frac{k}{\gamma} \simeq 4.7\frac{k}{\gamma}$$

giving an upper bound for the maximum height.

Another type of mechanism is represented in Figure V.7, where sliding with a velocity U occurs along an isolated straight line passing through the foot of the bank.

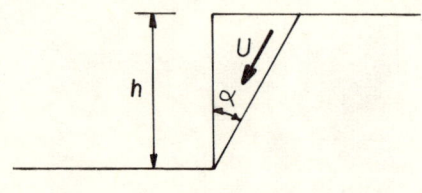

Figure V.7

The dissipation is equal to $k \times h/\cos \alpha \times U$ and the work done by the external forces is

$$\tfrac{1}{2}h^2 \tan \alpha \times \gamma \times U \cos \alpha$$

Hence, on balancing,

$$h = \frac{4k}{\gamma} \times \frac{1}{\sin 2\alpha}$$

giving an upper bound to the maximum height.

This mechanism can be optimized; i.e. it is possible to determine a value of α which provides the lowest upper bound. In fact, $\alpha = \pi/4$ and $h = 4k/\gamma$ gives a better upper bound for the limit height than the previous value.

Generally speaking, by searching for the best mechanism with an isolated line of sliding (circle), the upper bound is reduced to $3 \cdot 83 \, k/\gamma$. Josselin de Jong[1] and Heyman [25], by constructing an allowable stress-field, have proved that

$$h_1 \geqslant 2\sqrt{2}\frac{k}{\gamma}$$

Also, Palmer[2] has proved that

$$h_1 \geqslant 3\frac{k}{\gamma}$$

Hence

$$\frac{3k}{\gamma} \leqslant h_1 \leqslant 3.83\frac{k}{\gamma}$$

which appears to be the best bounds obtained by analytical means for the problem.

[1] 1965, unpublished.
[2] Private communication.

The exact value of h_1 for a bank made of a Tresca material is still unknown. For a bank made of a Tresca material with tension cut-off (i.e. no tensile stress allowed) the critical height is $h_1 = 2k/\gamma$ (see [17]).

9.2 Coulomb's wedge, Fellenius' method

The kinematic method with an isolated rectilinear line of sliding, as used above in the case of a Tresca material, has an equivalent in Soil Mechanics for the case of a Coulomb material.

The so-called 'Coulomb wedge' method is in fact a kinematic method, despite its static appearance. It uses an isolated, rectilinear line of velocity discontinuity for a standard Coulomb material [6]. It should be noted that along this line the velocity discontinuity is constant but not tangential (forming an angle ϕ), in accordance with Appendix A of Chapter IV (and also [17]).

Thus, Coulomb's wedge method, supplying an exterior approximation, will lead to an under-estimation of the active pressure and an over-estimation of the passive pressure in the case of a standard material, and *a fortiori* for a non-standard material.

Similarly, using the kinematic method with slip circles for a Coulomb material, gives solutions involving an isolated line of velocity discontinuity assuming the form of a logarithmic spiral. This is inclined at $(\pi/2 + \phi)$ to the vector radius coming from the rotational centre of the rigid block. The velocity discontinuity is not tangential to the line of discontinuity (which must be considered as a transition line in which dilatancy occurs).

The slip circle methods (originated by Fellenius) are often used in Soil Mechanics for the problems of slope stability. These methods cannot, for reasons given above, be interpreted as kinematic methods for a standard Coulomb material (the discontinuity velocity line not being a spiral), which explains their shortcomings in practical applications.

9.3 Finite element method

The theory of limit analysis makes it possible to use the finite element method, either in the static method for the construction of allowable stress fields [35, 36], or in the kinematic method for the construction of allowable velocity fields [20, 21]. The author believes that the kinematic approach is to be preferred to the static approach in problems involving large dimensions, e.g. stability of an embankment. It offers the possibility of limiting, *a priori*, the discretization to the regions known intuitively to be under greatest strain, the remainder of the system having a rigid-body motion determined by the boundary conditions.

The reader should refer to [20] for an example of the thrust of a smooth wall on a plastic infinite wedge of Tresca material. It is shown how the combination of static and kinematic methods and of the various processes for the construction of allowable fields, including the finite element method, can lead to remarkable results for the determination of the yield boundary of a system.

For the problem of the critical height of a bank[1] made of a Tresca material, using the finite element method in the static approach yields $3 \cdot 1\, k/\gamma$ as a lower bound for h_1 [43]. Various attempts made with finite elements through the kinematic method did not succeed in yielding a better upper bound for h_1 than $3 \cdot 83\, k/\gamma$.

Concerning three-dimensional continuous media, with the exception of axially symmetrical problems very few works are available dealing with the determination of limit loadings [18]. The finite method element could be very useful in this case.

References

[1] J. M. Alexander (1961) On complete solutions for frictionless extrusion in plane strain, *Quart. Appl. Math.*, **19**, No. 1, pp. 31–37.
[2] E. Anderheggen and H. Knöpfel (1972) Finite element limit analysis using linear programming, *Int. J. Solids & Structures*, **8**, 12, pp. 1413–1431.
[3] J. F. W. Bishop (1953) On the complete solution to problems of deformations of a plastic rigid material, *J. Mech. Sol.*, **2**, No. 1, pp. 43–53.
[4] S. J. Button (1953) The bearing capacity of footings on a two layer cohesive subsoil, *Proc. 3rd Int. Conf. Soil Mech. & Found. Engg.*, Zürich, **1**, pp. 332–335.
[5] I. F. Collins (1969) The upper bound theorem for rigid-plastic solids, *J. Mech. Phys. Sol.*, **17**, No. 5, pp. 323–338.
[6] I. F. Collins (1973) A note on the interpretation of Coulomb's analysis of the thrust on a rough retaining wall, *Géotechnique*, **23**, No. 3, pp. 442–447.
[7] J. Courbon (1971) *Plasticité Appliquée au Calcul des Structures*, E.N.P.C., Paris.
[8] M. Croc, G. Michel and J. Salençon (1971) Application de la programmation mathématique au calcul à la rupture des structures, *Int. J. Sol. Struct.*, **7**, No. 10, pp. 1317–1332.
[9] E. H. Davis and J. R. Booker (1971) The bearing capacity of strip footings from the stand-point of plasticity theory, *Univ. Sydney, Civ. Eng. Lab., Res. Rept.* R 170.
[10] E. H. Davis and J. R. Booker (1972) A note on a plasticity solution to the stability of slopes in inhomogeneous clays, *Géotechnique*, **22**, No. 3, pp. 509–513.
[11] E. H. Davis and J. R. Booker (1973) The effect of increasing shear strength on the bearing capacity of clays, *Géotechnique*, **23**, No. 4, pp. 551–563.
[12] L. Dietrich and W. Szczepinski (1969) A note on complete solutions for the plastic bending of notched bars, *J. Mech. Phys. Solids*, **17**, No. 3, pp. 171–176.
[13] A. Drescher (1971) A note on plane flow of granular media, *Problèmes de la Rhèologie, Symp. Franco-Polonais*, Varsovie 1971, pp. 135–144.
[14] A. Drescher (1972) Some remarks on plane flow of granular media, *Archives of Mechanics*, **24**, No. 5–6, pp. 837–848.
[15] D. C. Drucker (1954) Coulomb friction, plasticity and limit loads, *J. Appl. Mech.*, **21**, pp. 71–74.
[16] D. C. Drucker (1956) On uniqueness in the theory of plasticity, *Quart. Appl. Math.*, **26**, No. 1, pp. 35–42.
[17] D. C. Drucker and W. Prager (1952) Soil mechanics and plastic analysis or limit design, *Quart. Appl. Math.*, **10**, pp. 157–165.
[18] D. C. Drucker and R. T. Shield (1953) The application of limit analysis to punch-indentation problems, *Jl. Appl. Mech., Trans. ASME*, **20**, pp. 453–460.
[19] Y. d'Escatha and J. Mandel (1971) Profondeur critique d'éboulement d'un souterrain, *C.R. Ac. Sc., Paris, A*, **273**, pp. 470–473.

[1] It is to be remembered that the problem is considered as a plane strain problem.

[20] M. Fremond, A. Pecker and J. Salençon (1974) Méthode variationnelle pour le matériau rigide-plastique, *Symp. Franco-Polonais de Rhéologie et Mécanique des Sols*, Nice (Fr.), July 1974.

[21] M. Fremond and J. Salençon (1973) Limit analysis by finite element methods, *Symp. on the Role of Plasticity in Soil Mech.*, Cambridge (G.B.), Sept. 1973, pp. 297–308.

[22] J. P. Giroud and Tran Vo Nhiem (1971) Force portante d'une fondation sur une pente, *Ann. I.T.B.T.P.*, No. 283–284, pp. 130–180.

[23] J. P. Giroud, Tran Vo Nhiem and J. P. Obin (1974) *Tables pour le Calcul des Fondations*, t.3., Dunod, Paris.

[24] A. A. Gvozdev (1936) La détermination de la charge de ruine pour les systèmes hyperstatiques subissant des déformations plastiques, *Proc. Conf. on Plastic Deformations, Ac. Sc. USSR*, p. 19, v.z. transl. by R. M. Haythornthwaite in *Int. J. J. Mech. Sc.*, **1**, (1960), p. 322.

[25] J. Heyman (1973) The stability of a vertical cut, *Int. J. Mech. Sc.*, **15**, pp. 845–854.

[26] R. Hill (1948) A variational principle of maximum plastic work in classical plasticity, *Quart. J. Mech. Appl. Math.*, **1**, No. 18.

[27] R. Hill (1951) On the state of stress in a plastic-rigid body at the yield point, *Phil. Mag.*, **42**, p.. 868–875.

[28] R. Hill (1954) On the limit set by plastic yielding, *J. Mech. Sol.*, **2**, No. 4, pp. 278–285.

[29] P. G. Hodge (1970) Limit analysis with multiple load parameters, *Int. J. Solids & Structures*, **6**, pp. 661–675.

[30] D. D. Ivlev and R. I. Nepershin (1973) Impression of a smooth spherical indenter into a rigid-plastic half-plane, *Izv. AN. SSSR. Mekhanika Tverdogo Tela*, **8**, No. 4, pp. 159–166, Engl. Transl. in *Mechanics of Solids*, **8**, No. 4, pp. 144–149.

[31] G. de Josselin de Jong (1964) Lower bound collapse theorem and lack of normality of strain-rate to yield surface for soils, *Proc. IUTAM Symp. on Rheology & Soil Mechanics*, Grenoble (Fr.), pp. 69–75.

[32] G. de Josselin de Jong (1973) A limit theorem for material with internal friction, *Proc. Symp. on the Role of Plasticity in Soil Mechanics*, Cambridge (G.B.), pp. 12–21.

[33] G. de Kazinczy (1914) *Expéirneces sur les Poutres Encastrées*, Betonszemle, Vol. 2, p. 68.

[34] N. C. Kist (1917) Dissertation inaugurale. Ec. Polyt. Delft.

[35] J. Lysmer (1970) Limit analysis of plane problems in soil mechanics, *J. Soil Mech & Found. Div. ASCE*, **96**, No. SM4, Proc. Paper 7416, July 1970, pp. 1311–1334.

[36] G. Maier, A. Zavelani-Rossi and D. Benedetti (1972) A finite element approach to optimal design of plastic structures in plane stress, *Int. J. Num. Meth. Engng.*, **4**, pp. 455–473.

[37] J. Mandel (1966) *Mécanique des Milieux Continus*, Tome 2, Ann. 20, Gauthier-Villars, Paris.

[38] J. Mandel (1969) *Cours de Science des Matériaux*, Ecole Nationale Supérieure des Mines de Paris.

[39] J. Mandel and J. Salençon (1972) Force portante d'un sol sur une assise rigide (étude théorique), *Géotechnique*, **22**, No. 1, pp. 79–93.

[40] Ch. Massonet and M. Save (1970) *Calcul Plastique des Constructions*, Vol. 1 et 2, 2ème éd., Ed. CBLIA, Bruxelles.

[41] J. P. Obin (1972) Force portante en déformation plane d'un sol verticalement non homogène, Thèse Univ. Grenoble (Fr.).

[42] A. C. Palmer (1966) A limit theorem for materials with non associated flow laws, *J. Mécanique*, **5**, No. 2, pp. 217–222.

[43] J. Pastor (1976) Application de l'analyse limite à l'étude de la stabilité des pentes et des talus, Thèse Univ. Grenoble (France).

[44] D. Rádenkovic (1961) Théorèmes limites pour un matériau de Coulomb, *C.R. Ac. Sc. Paris*, **252**, pp. 4103–4104.

[45] D. Radenkovic (1962) Théorie des Charges Limites, in *Séminaire de Plasticité*, Ed. J. Mandel, P.S.T., Min. Air No. 116, pp. 129–142.
[46] G. Sacchi and M. Save (1968) A note on limit loads of non-standard materials, *Meccanica*, **3**, No. 1, pp. 43–45.
[47] J. Salençon (1969) La théorie des charges limites dans la résolution des problèmes de plasticité en déformation plane, Thèse Dr. Sc. Paris.
[48] J. Salençon (1972) Butée d'une paroi lisse sur un massif plastique: solutions statiques, *Jl. Mécanique*, **11**, No. 1, pp. 135–146.
[49] J. Salençon (1972) Un exemple de non validité de la théorie classique des charges limites pour un système non-standard, Proc. Int. Symp. on Foundations of Plasticity, Warsaw 1972, North Holland Publishing Co., Amsterdam.
[50] J. Salençon (1972) Théorie des charges limites, *Séminaire Plasticité et Viscoplasticité*, Eds. D. Rádenkovic and J. Salençon, Ediscience, Paris, 1974, pp. 207–229.
[51] J. Salençon (1972) Charge limite d'un systeme non-standard. Un contre-exemple pour la théorie classique, *Séminaire Plasticité et Viscoplasticité*, Eds. D. Radenkovic and J. Salençon, Ediscience, Paris, 1974, pp. 427–430.
[52] J. Salençon (1972) Ecoulement plastique libre et analyse limite pour les matériaux standards et non-standards, *Annales I.T.B.T.P.*, No. 295–296, July–August 1972, pp. 90–100.
[53] J. Salençon (1973) Prolongement des champs de Prandtl dans le cas du matériau de Coulomb, *Archives of Mechanics*, **25**, No. 4, pp. 643–648.
[54] J. Salençon (1974) Quelques résultats théoriques concernant la butée d'une paroi sur un coin plastique, *Annales I.T.B.T.P.*, No. 313, January 1974, pp. 185–194.
[55] J. Salençon (1974) Plasticité pour la Mécanique des Sols, C.I.S.M. Rankine Session, July 1974, Udine (It.).
[56] M. Save and Ch. Massonnet (1971) L'influence de la plasticité et de la viscosité sur la résistance et la déformation des constructions, *Rapp. Int. Congr. A.I.P.C.*
[57] M. Sayir and M. Ziegler (1968) Zum Prandtlschen Stampelproblem, *Ingenieur Archiv.*, **36**, No. 5, pp. 294–302.
[58] R. T. Shield (1954) Plastic potential theory and Prandtl bearing capacity solution, *Jl. Appl. Mech., transl. ASME*, **21**, pp. 193–194.
[59] W. Szczepinski (1966) Indentation of a plastic block by two opposite narrow punches, *Bull. Ac. Pol. Sc., série Sc. Tech.*, **14**, No. 11–12, pp. 671–676.

CHAPTER V

Appendixes

A. AN ALTERNATIVE PRESENTATION OF THE THEORY OF LIMIT ANALYSIS. EXTENSION TO SOME NON-STANDARD MATERIALS

1. Introduction

This presentation is not intended to be a formal development of the theory of limit analysis, but attempts, by adopting a more mathematical mode of exposition, to present more clearly the hypotheses required for each result. This will make it easier to pass on to non-standard materials, and may increase understanding between numerical analysts and engineers. A very similar presentation has already been adopted in [12, 14].

2. The Case of a Standard Material

2.1 Theorem 1

Let f be a convex loading function. The surface $f(\sigma) = 0$ is assumed to be 'open' in several possible directions which are all necessarily, as a consequence of the convexity of f, the directions of a convex cone.

Then, the set

$$G' = \{\mathbf{v} | \mathbf{v} \in \lambda \partial f(\sigma), \lambda \geq 0, f(\sigma) = 0, |\sigma| < \infty\} \tag{1}$$

is a convex cone with a summit O of R^6. This cone is the complementary of the cone of directions in which the surface $f(\sigma) = 0$ is open.

Mechanical aspect:

For a standard material with a loading function f, G' is the convex cone of the plastically admissible strain rate tensors. Usually, if the loading surface is 'open' in the one direction of the isotropic pressures, then G' is a half-space, or, if the loading function is 'open' in the directions of isotropic pressures and tensions, it is necessarily cylindrical, and G' is a plane.

2.2 Theorem 2. Dissipation

In all that follows, f is a convex loading function, and it is assumed that $f(0) \leq 0$.

The bilinear form is considered:

$$F(\sigma, v) = \sigma \cdot v.^{1} \qquad (2)$$

where v is a strain-rate tensor, and σ is a stress tensor. The convex set,

$$G = \{\sigma \mid f(\sigma) \leq 0\} \qquad (3)$$

is used to look for

$$\text{Sup}\{F(\sigma, v) \, \sigma \in G\} \qquad (4)$$

(1). If $v \in G'$ then $F(\sigma, v)$ has a finite maximum, reached for

$$\sigma: f(\sigma) = 0, v \in \lambda \partial f(\sigma), \lambda > 0 \qquad (5)$$

The value of this maximum will be denoted by

$$\pi(v) = \text{Sup}\{F(\sigma, v) \mid \sigma \in G\}$$

Since $\sigma = 0 \in G$, $\pi(v) \geq 0$.

(2). If $v \notin G'$

$$\text{Sup } F(\sigma, v) = +\infty$$

In this case,

$$\pi(v) = \text{Sup}\{F(\sigma_1, v) \mid \sigma \in G\} = +\infty$$

Mechanical aspect:

(1). For a material with a loading function f, G is the convex domain of plastically admissible stress tensors.

(2). For a standard material with a convex loading function f, $\pi(v)$, in the case where $v \in G'$, is the dissipation for the strain-rate tensor v. Equation (5) corresponds to the inversion of the flow rule for v.

2.3 Theorem 3. Convexity of π

$$v^1, v^2 \in G' \text{ are given.}$$

Consider

$$v = \lambda v^1 + (1 - \lambda) v^2, \lambda \in [0, 1]$$

According to Theorem 1, $v \in G^1$

Let σ be a stress tensor associated with v by (5), so that

$$\pi(v) = \sigma \cdot v = \lambda \sigma \cdot v^1 + (1 - \lambda) \sigma \cdot v^2$$

---

[1] $\sigma \cdot v = \sigma_{ij} v_{ij}$

From the definitions of $\pi(\mathbf{v}^1)$ and $\pi(\mathbf{v}^2)$

$$\boldsymbol{\sigma} \cdot \mathbf{v}^1 \leq \pi(\mathbf{v}^1)$$
$$\boldsymbol{\sigma} \cdot \mathbf{v}^2 \leq \pi(\mathbf{v}^2)$$

whence,

$$\pi(\mathbf{v}) \leq \lambda \pi(\mathbf{v}^1) + (1 - \lambda) \pi(\mathbf{v}^2) \tag{6}$$

and π is, therefore, a convex function on G'.

According to the adopted convention,

$$\pi(\mathbf{v}) = +\infty \quad \text{if} \quad \mathbf{v} \notin G'$$

and π is convex on the set of all strain-rate tensors (R^6).

Another property is that

$$\pi(\lambda \mathbf{v}) = \lambda \pi(\mathbf{v}) \,\forall \lambda \geq 0$$

and π is a one-degree positive homogeneous function.

2.4 Definitions

The symbols σ and v denote stress- and strain-rate fields ($\boldsymbol{\sigma}$ and $\mathbf{v}$ denote the tensors, i.e. values of these fields at a particular point.)

It is not intended, in the following, to describe from a mathematical viewpoint the appropriate functional spaces in which the specified integrals have their meaning.[1]

The first definition used is that

$$H = \{\sigma \,|\, \boldsymbol{\sigma} \in G; \exists J_d \in \mathcal{D}_p \text{ such that } \sigma \text{ be S.A. ass. } J_d\}$$

THEOREM: H forms a convex surface.[1]

As $\mathcal{D}_p$ is assumed to be a linear space (see Appendix of Chapter III), and furthermore, G is a convex set, it is immediately verified that, if $\sigma' \in H$, $\sigma^2 \in H$,

$$\lambda \sigma' + (1 - \lambda)\sigma^2 \in H, \qquad \lambda \in [0, 1].$$

It should be noted that since $f(\mathbf{0}) \leq 0$, H contains 0.

The second definition is that

$$H' = \{v \,|\, \mathbf{v} \in G'; \exists J_c \in \mathcal{C}_p \text{ such that } v \text{ be K.A. ass. } J_c\}$$

THEOREM: H' is a convex cone with a summit 0.

It suffices to state that $\mathcal{C}_p$ is a linear space and G' is a convex cone with a summit 0.

Mechanical aspect:

H is the convex set of the allowable stress-fields (statically and plastically

---

[1] It is understood that the material constituting the system is not necessarily homogeneous. At each point, suitable G and G' spaces will be considered.

admissible) for the system under study, subject to a loading process with n parameters.

H' is the cone of the allowable velocity fields (kinematically and plastically admissible) for the system under study in the same conditions.

2.5 Functional I

A functional I is defined by

$$I(\sigma, v) = \int_V \pi(\mathbf{v}) \, dV - \mathbf{Q}(\sigma) \cdot \dot{\mathbf{q}}(v) = \int_V [\pi(\mathbf{v}) - \boldsymbol{\sigma} \cdot \mathbf{v}] \, dv \qquad (7)$$

for any statically admissible stress field σ and any kinematically admissible strain rate field v.

This functional is non-negative on $H \times H'$, i.e.:

$$\forall \sigma \in H, \quad \forall v \in H',$$

we have

$$I(\sigma, v) \geqslant 0 \qquad (8)$$

This result follows from (7) and the definition of $\pi(\mathbf{v})$.

2.6 Theorem

THEOREM: A solution to the problem of uncontained plastic flow for a system of a standard material renders $I(\sigma, v)$ a minimum on $H \times H'$.

Proof:
Such a solution consists of:

(1). An allowable stress-field, $\sigma \in H$, and
(2). An allowable strain rate field, $v \in H'$.

Both fields are associated by the constitutive law for a standard material; i.e. if $\mathbf{v} \neq 0$, then $f(\boldsymbol{\sigma}) = 0$ and $\mathbf{v} \in \lambda \partial f(\boldsymbol{\sigma})$, $\lambda > 0$. It follows that for both fields

$$I(\sigma, v) = \int_V [\pi(\mathbf{v}) - \boldsymbol{\sigma} \cdot \mathbf{v}] \, dV = 0 \qquad (9)$$

Therefore, I is a minimum according to (8).

2.7 Converse

THEOREM: Any solution (σ, v) to the problem

$$\text{Min } I(\sigma, v) \qquad (\sigma, v) \in H \times H' \qquad (10)$$

such that $v \neq 0$, is a solution of the problem of uncontained flow for the system of a standard material.

Proof:
It is known that

$$\text{Min } I(\sigma, v) = 0, \qquad (\sigma, v) \in H \times H'$$

as the field $v = 0$ belongs to H'. Therefore, any other solution of the minimization problem is such that

$$I(\sigma, v) = 0$$

At any point where $\mathbf{v} \neq 0$ it follows, according to the definition of π,

$$\pi(\mathbf{v}) = \boldsymbol{\sigma} \cdot \mathbf{v}$$

that $f(\sigma) = 0$ and $\mathbf{v} \in \lambda \partial f(\boldsymbol{\sigma})$, $\qquad \lambda > 0$.

A solution to the problem of uncontained plastic flow for the system of a standard material is, therefore, obtained.

2.8 Other definitions

Let the set K be defined by:

$$\text{When } \sigma \text{ describes } H, \mathbf{Q}(\sigma) \text{ describes } K \subset R^n$$

It is immediately verified that K is a convex set.
Let the K' be defined by:

$$\text{when } v \text{ describes } H', \dot{\mathbf{q}}(v) \text{ describes } K' \subset R^n$$

It is immediately verified that K' is a convex cone with a summit 0.

Mechanical aspect:
K is the convex set of the allowable loadings of the system. K' is the convex cone of the allowable strain rates of the system.

2.9 Corollary 1: theorem of maximum work

Let (σ, v) be a solution to the problem of uncontained flow for a standard material.
Let $\sigma' \in H$, so that

$$I(\sigma', v) \geqslant I(\sigma, v)$$

or,

$$\int_V \pi(\mathbf{v}) \, dV - \mathbf{Q}(\sigma') \cdot \dot{\mathbf{q}}(v) \geqslant \int_V \pi(\mathbf{v}) \, dV - \mathbf{Q}(\sigma) \dot{\mathbf{q}}(v)$$

whence

$$[\mathbf{Q}(\sigma) - \mathbf{Q}(\sigma')] \cdot \dot{\mathbf{q}}(v) \geqslant 0 \tag{11}$$

As a consequence, the $\mathbf{Q}(\sigma)$ limit loading belongs to the boundary of K and $\dot{\mathbf{q}}(v)$ is an outward normal to this boundary.

Mechanical aspect:

The inequality (11) expresses the theorem of maximum work (Section V.3.5) from which the static theorem and the static method may be derived (Sections V.3.8 and 3.9.).[1]

2.10 Corollary 2: convex K_1, concave L_1

2.10.1 *Convex K_1*

The convex set K_1 in space $\{Q\} = R^n$ is defined by

$$K_1 = \bigcap_{v \in H'} \left(\mathbf{Q} \,\Big|\, \int_V \pi(\mathbf{v})\,dV - Q \cdot \dot{\mathbf{q}}(v) \geqslant 0 \right) \qquad (12)$$

It follows from Section 2.2 that the same set K_1 is obtained by considering in (12) all the K.A. strain-rate fields, whether or not they are P.A.

Considering now any $\mathbf{Q}' \in K$ and a corresponding allowable stress field $\sigma' \in H$, it follows from (8) that:

$$\forall v \in H' \qquad I(\sigma', v) = \int_V \pi(\mathbf{v})\,dV - \mathbf{Q}'\dot{\mathbf{q}}(v) \geqslant 0$$

Therefore
$$\mathbf{Q}' \in K_1$$

Whence
$$K \subset K_1$$

Consequently, given $v \in H'$, the half-space

$$\int_V \pi(\mathbf{v})\,dV - \mathbf{Q}\dot{\mathbf{q}}(v) \geqslant 0$$

contains the set K_1, and, therefore, also the set K.

Mechanical aspect:

The result obtained is the basis of the kinematic theorem and the kinematic method (Section V.4.3).[1]

2.10.2. *Concave L_1*

L_1 is concave since its complementary set in space $\{Q\}$ is evidently convex. Moreover
$$CK_1 \subset L_1 \subset \overline{C}K_1$$

K_1 and L_1 have the same boundary. As π is a convex function of $\mathbf{v} \in R^6$, it follows that $\int_V \pi(\mathbf{v})\,dV$ is a convex function or v K.A. This entails that

$$\operatorname{Inf}\left\{ \int_V \pi(\mathbf{v})\,dV \,\Big|\, v \text{ K.A., } \dot{\mathbf{q}}(v) = \dot{\mathbf{q}} \right\}$$

---

[1] Without making use of the existence hypothesis as indicated in Section V.5.1.

is a convex function of $\dot{\mathbf{q}} \in R^n$. This function is the supporting function of K_1, meaning that any plane tangential to the K_1 (and L_1) boundary assumes the form

$$\mathrm{Inf}\left\{\int_V \pi(\mathbf{v})\,dV \,\middle|\, v \text{ K.A.}, \dot{\mathbf{q}}(v) = \dot{\mathbf{q}}\right\} - \mathbf{Q}\dot{\mathbf{q}} = 0, \qquad \dot{\mathbf{q}} \in R^n.$$

2.11 Properties of the limit loadings

If Q_1 is a limit loading, then

(1). K is not open at point $\mathbf{Q}_1$,
(2). $\mathbf{Q}_1 \in L_1$, which is not open at this point,
(3). K and K_1 are tangential in $\mathbf{Q}_1$.

In fact $\mathbf{Q}_1 \in K$ and, according to Section 2.9., $\mathbf{Q}_1$ is on the boundary of K.
On the other hand, if v is a strain-rate field corresponding to this limit loading, it follows (Section 2.6) that

$$\int \pi(\mathbf{v})\,dV - \mathbf{Q}_1 \dot{\mathbf{q}}(v) = 0$$

and, therefore,

$$\mathbf{Q}_1 \in L_1.$$

It results from Section 2.10.2 that at this point L_1 cannot be open and Q_1 belongs to the L_1 (and K_1) boundary.

Furthermore, the plane

$$\int \pi(\mathbf{v})\,dV - \mathbf{Q}\dot{\mathbf{q}}(v) = 0 \tag{13}$$

passes through $\mathbf{Q}_1$, and is a tangent to K (since $\dot{\mathbf{q}}(v)$ is an outward normal to K). It cannot intersect K_1, by definition; therefore, it is a tangent to K_1 also.

2.12 Existence hypothesis (of the solution to the problem of uncontained plastic flow)

The hypothesis assumes that

(1). $K \equiv K_1$ (as a consequence, K is closed),
(2). $L_1 = CK_1$ (stated otherwise, L_1 is closed).

Mechanical aspect:

This hypothesis is equivalent to the following proposition, which has been presented as the existence hypothesis in Section V.3.7.

For a proportional loading process given by

$$\mathbf{Q} = \lambda \mathbf{Q}^*, \qquad \lambda > 0$$

either the loading can increase endlessly, i.e. λQ^* is allowable, $\forall \lambda > 0$, or there exists a limit loading $\mathbf{Q} = \lambda Q^*$ corresponding to the solution of a problem of uncontained plastic flow.

Effectively, either λQ^* is the direction of a point of K at infinity, or there exists in this direction a point $\mathbf{Q}_1$ on the common boundary of the closed sets $K, K_1,$ and L_1. In this case,

$$\mathbf{Q}_1 \in K \Rightarrow \exists \sigma_1 \in H: \quad \mathbf{Q}_1 = \mathbf{Q}(\sigma_1)$$

$$\mathbf{Q}_1 \in K_1 \quad \text{and} \quad \mathbf{Q}_1 \in L_1 \Rightarrow \exists v_1 \in H', \quad v_1 \neq 0:$$

$$\int \pi(\mathbf{v}_1) \, dV - \mathbf{Q}_1 \dot{\mathbf{q}}(\mathbf{v}_1) = 0$$

(giving properties of the planes which are tangential to the boundary of the closed set L_1), or

$$I(\sigma_1, v_1) = 0$$

so that (σ_1, v_1) is a solution of the problem of uncontained plastic flow. The proof of the reciprocal property is derived from Section 2.6.

It follows from this result that any point at a finite distance, of the common boundary of K, K_1 and L_1, is a limit loading.

Remark:

The necessity of the existence hypothesis may appear merely mathematical since, when stated in a mechanical form it seems physically self-evident. But it is to be remembered that the concept of a rigid-plastic material is purely mathematical by itself; an example has been recently given [17] of a problem where a limit loading does exist without uncontained plastic flow being possible.

2.13

Let $\mathbf{Q}^*$ be a loading such that
$$\lambda Q^* \in K, \qquad \forall \lambda \geq 0.$$

Let K.A. ass. with $J_c \in C_p$ be such that

$$\mathbf{Q}^* \cdot \dot{q}(v) > 0 \tag{14}$$

Then, either $v \notin H'$, or $v \in H'$, and the plastically deformed zone in this field has an infinite extent.[1]

Proof:

According to Section 2.10, $K \subset K_1$, so that (12) is verified by λQ^*, and as v is kinematically admissible, allowable or not, it follows that

$$\int_V \pi(\mathbf{v}) \, dV - \lambda \mathbf{Q}^* \dot{\mathbf{q}}(v) \geq 0 \quad \forall \lambda > 0 \tag{15}$$

---

[1] In practical problems, this circumstance will often be excluded by the very form of the boundary conditions: generally, $u = 0$ at ∞; while $T = 0$ at ∞ is much rarer.

Hence, using (14),

$$\int_V \pi(\mathbf{v})\,dV = +\infty$$

which proves the predicted result.

2.14 Consequences of the existence hypothesis; practical methods; theorems

2.14.1 *Static method*

The yield boundary is often determined with the help of the static method by considering a loading process where the values of Q_i vary proportionally.

Let $\mathbf{Q}^d$ be a given loading. The uncontained plastic flow solutions corresponding to limit loadings assuming the form $\mathbf{Q} = \lambda \mathbf{Q}^d$ (possibly with the restriction $\lambda > 0$) are now sought. This requires solving the problem

$$\text{Min}\,\{I(\sigma, v) \,|\, v \in H'; \sigma \in H, \mathbf{Q}(\sigma) = \lambda \mathbf{Q}^d\} \tag{16}$$

and keeping only the solutions in which $v \neq 0$.

Using the notation

$$\varepsilon_v = \sin(\mathbf{Q}^d \dot{\mathbf{q}}(v))$$

expression (16) can be written as

$$\text{Min}\,\left\{\int_V \pi(\mathbf{v})\,dV - \varepsilon_v \dot{\mathbf{q}}(v)\mathbf{Q}^d \cdot \text{Max}\,\{\varepsilon_v \lambda(\sigma) \,|\, \sigma \in H, \mathbf{Q}(\sigma) = \lambda(\sigma)\mathbf{Q}^d\} \,\Big|\, v \in H'\right\} \tag{17}$$

Solving the problem

$$\text{Sup}\,\{\varepsilon_v \lambda(\sigma)\}$$

separately for the cases $\varepsilon_v = +1$, and $\varepsilon_v = -1$, gives two values λ^1_+ and λ^1_- (resp. >0, <0).

The range for σ in (17) is a convex set $\subset H$, on which are found two values λ^1_+ and λ^1_- (resp. >0, <0) corresponding to $\text{Sup}\,\{\varepsilon_v \lambda(\sigma)\}$ according to the field v. The loadings $\lambda^1_+ \mathbf{Q}_d$ and $\lambda^1_- \mathbf{Q}_d$ are on K's boundary at the intersection with the straight line $\mathbf{Q} = \lambda \mathbf{Q}^d$ (Figure V.A.1).

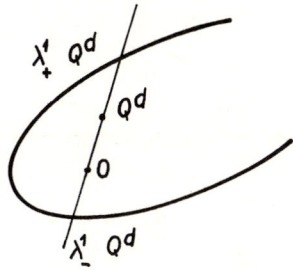

Figure V.A.1

As a consequence of the existence hypothesis the loadings dealt with are limit loadings (if they are at a finite distance). This means that they correspond to fields σ_+^1 (resp. σ_-^1) of H which realize the Max term in expression (17) (Min in σ in (16)).

The existence of a solution $v'_+ \neq 0$ of the problem

$$\text{Min}\left\{\int_V \pi(\mathbf{v})\,dV - \dot{\mathbf{q}}(v)\mathbf{Q}^d\lambda_+^1 \,\big|\, v \in H',\, \varepsilon_v = +1\right\} \tag{18}$$

is ascertained. The solution of (18) will be useless if, as is the general case, only the limit loadings are required.

The practical static method involves determining *the limit loading by solving the problem*

$$\text{Max}\,\{\lambda(\sigma)\,|\,\sigma \in H,\, \mathbf{Q}(\sigma) = \lambda(\sigma)\mathbf{Q}^d,\, \lambda(\sigma) > 0\} \tag{19}$$

Thus, an attempt is made to investigate the convex set of (19), so as to maximize λ. K is explored, following vector radii, and by dealing with a sub-set of the convex set of (19), *an approximation of the valid boundary from inside* is obtained.

2.14.2 Kinematic method

The kinematic determination of the yield boundary is effected by setting $\dot{\mathbf{q}}(v)$ in (12).

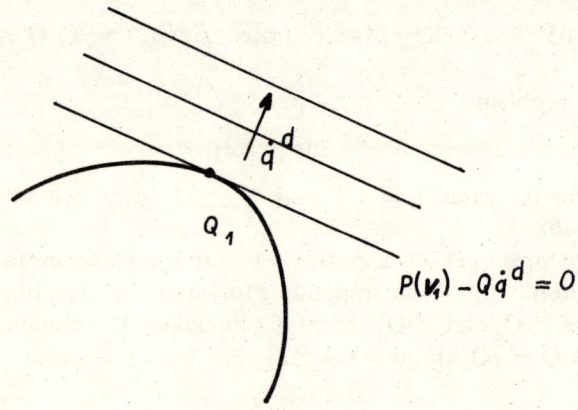

Figure V.A.2

Let $\dot{\mathbf{q}}^d \in K'$, $\dot{\mathbf{q}}^d \neq 0$. The solutions of the uncontained flow problem are now sought, such that the corresponding strain rate of the system (Figure V.A.2) be given by

$$\dot{\mathbf{q}}(v) = \dot{\mathbf{q}}^d$$

Thus, the solution of the problem

$$\text{Min}\,\{I(\sigma, v)\,|\,\sigma \in H;\, v \in H',\, \dot{\mathbf{q}}(v) = \dot{\mathbf{q}}^d\} \tag{20}$$

is a solution to the uncontained flow problem if Min is zero. Using the notation

$$P(v) = \int_V \pi(\mathbf{v}) \, dV$$

equation (20) may be written as

$$\text{Min}\{P(v)|v \in H', \dot{\mathbf{q}}(v) = \dot{\mathbf{q}}^d\} - \text{Max}\{\mathbf{Q}(\sigma)\dot{\mathbf{q}}^d|\sigma \in H\}. \tag{21}$$

Here, v is obtained independently of σ by solving the Min problem in (21) and σ independently of v by solving the Max problem.

The existence of $\text{Sup}\{Q(\sigma)\dot{\mathbf{q}}^d|\sigma \in H\}$ is ensured and the corresponding loading $\mathbf{Q}_1$ is the point of K's boundary where $\dot{\mathbf{q}}^d$ is an outward normal. According to the existence hypothesis this is a limit loading. Therefore, it corresponds to a stress-field $\sigma_1 \in H$ which realizes the Max in (21); and there exists $v_1 \neq 0$ which realizes the Min in (21),[1] such that

$$I(\sigma_1, v_1) = 0$$

Stated otherwise, the plane

$$P(v_1) - \mathbf{Q}\dot{\mathbf{q}}^d = 0 \tag{22}$$

passes by $\mathbf{Q}_1$ and is a tangent to the yield boundary. The classical kinematic methods is subsequently derived. *The half-space*

$$\text{Min}\{P(v)|v \in H', \dot{\mathbf{q}}(v) = \dot{\mathbf{q}}^d\} - \mathbf{Q}\dot{\mathbf{q}}^d \geq 0 \tag{23}$$

contains K and is a tangent to the yield boundary.

The convex set of (23) is explored so as to minimize $P(v)$. Thus, L_1 is explored by means of half-spaces of fixed direction. In practice, a sub-set of the convex set of (23) is used to obtain an *approximation of the yield boundary from outside*.

Thus, a static dual approach, also, is available.
The loading obtained by solving the problem

$$\text{Max}\{\mathbf{Q}(\sigma)\dot{\mathbf{q}}^d|\sigma \in H\}, \qquad \dot{\mathbf{q}}^d \in K'$$

is a limit loading for which $\dot{\mathbf{q}} = \dot{\mathbf{q}}^d$.

Finally, the condition $\dot{\mathbf{q}}^d \in K'$ stated at the outset prevents the convex set of (23) from being void, but it is sometimes difficult to know, *a priori*, whether $\dot{\mathbf{q}}^d \in K'$. In fact, in (20) and also in the statements of the classical kinematic method and its dual approach, any $\dot{\mathbf{q}}^d$ whatever in R^n may be considered, both latter problems being stated

$$\text{Inf}\{P(v)|v \text{ K.A.}, \dot{\mathbf{q}}(v) = \dot{\mathbf{q}}^d\} - \mathbf{Q}\dot{\mathbf{q}}^d \geq 0 \tag{23'}$$

$\text{Sup}\{\mathbf{Q}(\sigma)\dot{\mathbf{q}}^d|\sigma \in H\}$, for any $\dot{\mathbf{q}}^d$ whatever, and the results will be unchanged as, (1). If

$$\dot{\mathbf{q}}^d \in K', \quad P(v) \equiv +\infty$$

---

[1] Effectively, $\dot{\mathbf{q}}^d$ is an outward normal at $\mathbf{Q}_1$ to the boundary of L_1, which is closed.

and, therefore,

$$\text{Inf}\{P(v)\} = +\infty, \quad (\Leftrightarrow \text{Sup}\{\mathbf{Q}(\sigma)\dot{\mathbf{q}}^d\} = +\infty$$

by application of the existence hypothesis, and there is no limit loading.

(2). If

$$\dot{\mathbf{q}}^d \in K' \quad \text{and} \quad \text{Inf}\{P(v)\} < +\infty$$

a Min is reached for $v \in H'$ (as $v \notin H' \Rightarrow P(v) = +\infty$) and again the initial formulation results.

Lastly, the absence of a solution can occur, even if $\dot{\mathbf{q}}^d \in K$, $\text{Inf}\{P(v)\} = +\infty$ can be realized in this case, when all the fields v of H', such that $\dot{\mathbf{q}}(v) = \dot{\mathbf{q}}^d$, have an unlimited deformed zone (this agreeing with the results of Section 2.13).

2.14.3 *Theorem of association*

When using the static and kinematic method, it is assumed that firstly a stress-field $\sigma^* \in H$ is found and $\mathbf{Q}^*$ is made equal to $\mathbf{Q}(\sigma^*)$, and secondly a strain-rate field $v \in H'$ with $\dot{\mathbf{q}}^*$ equal to $\dot{\mathbf{q}}(v^*)$, such that

$$P(v^*) - \mathbf{Q}^* \cdot \dot{\mathbf{q}}^* = 0$$

The following can then be stated.

(1). $\mathbf{Q}^*$ is a limit loading.

(2). $\dot{\mathbf{q}}^*$ is a strain rate of the system corresponding to this loading.

(3). The fields σ^* and v^* form a limit equilibrium solution corresponding to $\mathbf{Q}^*, \dot{\mathbf{q}}^*$.

The proof of this proposition is immediate, since $I(\sigma^*, v^*) = 0$.

The uniqueness theorem of the stress field (for certain assumptions, in the union of the deformed zones) presented in Chapter III, develops as a consequence of the theorem of association (Figure V.A.3).

2.14.4 *The case of a loading process with one parameter*

In the frequent case of a loading process depending on one parameter $Q\ (>0)$,

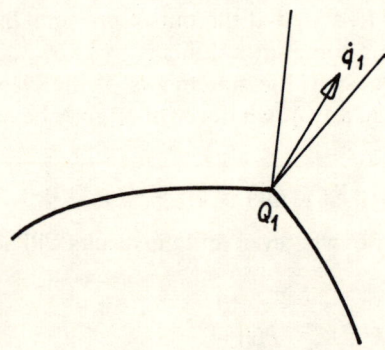

Figure V.A.3

the statements corresponding to Sections 2.14.1 and 2.14.2 assume simple classical forms.

Static method:
$$Q_{\lim}^+ = \operatorname{Max}\{Q(\sigma)|\sigma \in H\}$$

Kinematic method:
$$Q_{\lim}^+ = \operatorname{Min}\{P(v)|v \in H', \dot{q}(v) = 1\}$$

and its dual formulation,
$$Q_{\lim}^+ = \operatorname{Max}\{Q(\sigma)|\sigma \in H\}$$

which is identical to that of the usual static method.

3. Case of Non-standard Systems

3.1 Definition

The classical theory of limit analysis presented in the preceding paragraph explicitly uses both properties of convexity and normality, equivalent to the principle of maximum plastic work.

As already stated in Chapter I, the convexity of the yield criterion is a property with almost general validity. On the contrary, the normality of the flow rule is often non-existent. In particular, this is the case for soils obeying the Coulomb criterion, and also, as will be seen in Section 4, in many cases for the condition of contact at the interfaces between the solids which constitute a system.

Thus, the case dealt with will be that of systems with a convex yield criterion,[1] but with a flow rule which does not satisfy the condition of normality at any point. These are *non-standard systems*.

This type of system was first investigated by Drucker [3] with regard to the conditions at the interfaces; then followed a general survey by Radenkovic [7, 8] and studies by Josselin de Jong [4, 5], Palmer [6], Collins [1, 2] and Salençon [11, 12, 16].

3.2 A kinematic method

The symbol M denotes the constitutive material[2] of the non-standard material (M). The yield criterion of M (assumed to be convex) is represented by f. The definition of a plastically admissible stress tensor for M is obviously unchanged with respect to the case of a standard material:
$$f(\boldsymbol{\sigma}) \leqslant 0$$
and hence the set
$$G_M = \{\boldsymbol{\sigma}|f(\boldsymbol{\sigma}) \leqslant 0\}$$

---

[1] The expression 'yield criterion' is taken in its widest sense: it includes not only the yield criterion of the constitutive materials, but also the friction condition at the interfaces between the solids that constitute a system.

[2] M includes the constitutive material itself and the contact interface between solids of the system.

Similarly defined is the set of allowable stress fields (or S.P.A.) for (M),

$$H_M = \{\sigma \,|\, \sigma \text{ S.A.}, \sigma \in G_M \text{ everywhere}\}$$

and the set of allowable loadings for (M),

$$K_M = \{\mathbf{Q}(\sigma) \,|\, \sigma \in H_M\}$$

As a consequence of the convexity of f, the sets G_M, H_M, K_M are *convex*, following the same arguments as in Section 2. It results from the definition of K_M that all the loadings that can be borne by (M) are necessarily included, and, in particular, the limit loadings for (M).

If F denotes a standard material with a yield criterion f, and (F) a system geometrically similar to (M) and constituted by F, the sets G_M, H_M, K_M are identical to spaces G, H, K, as defined for (F) in Section 2, and will be denoted by G_F, H_F, K_F. In particular,

$$K_M = K_F$$

Whereas for (F) the boundary of K_F is the yield boundary, the same property cannot be proved for (M), for the normality rule states that the standard system (F) 'proceeds to the limit of its possibilities'.

The boundary of K_M appears for (M) as the boundary of tolerable loads, and, in particular, as an outer boundary of limit loadings. Its determination is interesting for this reason. As a consequence of the identity $K_M = K_F$, this determination is carried out through the intermediary of the standard system (F), *as an auxiliary for the calculation* by either

(1). The static method for (F), or
(2). The kinematic method for (F).

This supplies an approximation of the boundary of K_M from the outside, which in itself is an outer boundary of the limit loadings for (M), and justifies the title of *kinematic method for* (M) (Figure V.A.4).

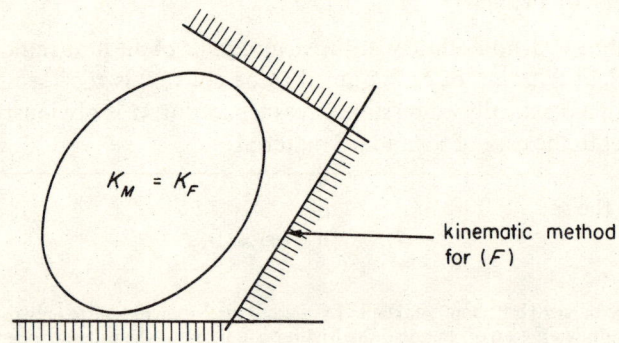

Figure V.A.4

3.3 Radenkovic theorem—a static method

3.3.1 *Type of non-standard systems studied*

The non-standard system (*M*) considered has a flow rule with the following properties at each point.

There exists a function g, convex on R^6, such that

$$g(0) \leq 0$$

and

$$\left.\begin{array}{c} \forall \sigma \text{ verifying } f(\sigma) = 0, \\ \exists \sigma'(\sigma) \text{ verifying } g(\sigma') = 0 \\ \mathbf{v}^P(\sigma) \in \lambda \partial g(\sigma'), \quad \lambda \geq 0 \end{array}\right\} \quad (24)$$

Also

$$(\sigma - \sigma') \cdot \mathbf{v}^P(\sigma) \geq 0 \quad (25)$$

3.3.2 *Remarks*

(1). Standard systems are obviously a particular case of the non-standard systems presented above, and for these $f = g$.

(2). A Coulomb material, with a flow rule defined by a constant dilatancy angle v, satisfies the conditions of Section 3.3.1. The value of g is given by the Coulomb loading function with an angle v and a cohesive pressure equal to that of f: $cg/\tan v = c/\tan \phi$.

(3). For a non-standard material satisfying the conditions of Section 3.3.1, there is no uniqueness of function g or of the correspondence

$$\sigma \rightarrow \sigma'(\sigma)$$

Thus, if g and $\sigma \rightarrow \sigma'$ satisfy the conditions, then g_m defined by $g_m(\alpha) = g(m\alpha) \forall \alpha$, with $m > 1$ and $\sigma \rightarrow \sigma'' = \sigma'/m$, also satisfy them.

Similarly, some translations can be effected on σ' and on g, etc.

(4). It results from (25) and the convexity of f and g that if G_G is defined by

$$G_G = \{\sigma | g(\sigma) \leq 0\}$$

then

$$G_G \subset G_F$$

(5). In Chapter I the concept of a plastic potential that was different from the yield criterion was evoked, $\mathbf{v}^P$ appearing as a sub-gradient of a function of σ. This function is not usually the function g that appears in Section 3.3.1.

In the case of a non-standard Coulomb material, with a flow rule defined by a constant dilatancy angle v ($0 \leq v \leq \phi$) or with a standard von Mises flow rule, the function g and the plastic potential can become the same. It was from this case [7] that the concept of a plastic potential different from the yield criterion was first generalized [8]. The important concept in the general case, for what follows, is that of the function g.

3.3.3 System (G)—Radenkovic theorem

For a material M of the type indicated in Section 3.3.1, and for each function g, an associated material G is defined. This is the standard material with a criterion g. The system (G) is that constituted by G, which is geometrically similar to (M).

A solution of the problem of uncontained plastic flow for (M), constituted by the fields σ and v, gives

$$\sigma \in H_F \Rightarrow \mathbf{Q}(\sigma) \in K_F = K_M \quad \text{(Section 3.2)}$$

and as a consequence of (24), $v \in H'_G$, the set of the all allowable velocity fields for (G). Hence, $\dot{\mathbf{q}}(v) \in K'_G$, the set of all allowable strain rates of (G).

For the field σ' defined at any point of the zone formed by (24), it follows from (25) that

$$\int_V (\boldsymbol{\sigma} - \boldsymbol{\sigma}') \cdot \mathbf{v} \, dV \geqslant 0 \tag{26}$$

or, also,

$$\int_V \boldsymbol{\sigma}' \cdot \mathbf{v} \, dV - \mathbf{Q}(\sigma) \cdot \dot{\mathbf{q}}(v) \leqslant 0 \tag{27}$$

The first term of (27) is the dissipation $P_G(v)$ of field v for the system (G), and thus

$$P_G(v) - \mathbf{Q}(\sigma)\dot{\mathbf{q}}(v) \leqslant 0 \tag{28}$$

which proves, by application of the results of Section 2.10.2, that $\mathbf{Q}(\sigma)$ *is not internal to* K_G, *the set of allowable loadings for* (G).[1]

3.3.4 Consequences

The boundary of K_G, therefore, appears as an inner boundary for the limit loadings of the system (M). Its determination will be carried out by either the kinematic or the static method of the classical theory, as (G) is standard. The static method leads to an approximation from the inside of the boundary of K_G, justifying the name of *static method for* (M) (see Figure V.A.5).

Thus, by grouping the results of Sections 3.2 and 3.3, for the conditions of validity of the Radenkovic theorem, it is proved that the limit loadings are to be found in the 'ring' comprised between K_F and K_G, the boundaries being included:

$$\mathbf{Q}_{\text{lim}} \in \overline{K_F - K_G} \tag{29}$$

---

[1] The importance of this result should be emphasized. In a solution (σ, v) of the problem of uncontained flow for (M), the yield criterion $f = 0$ is necessarily reached at some points. Thus, σ is on the boundary of H_F. Now $G_G \subset G_F$ and therefore $H_G \subset H_F$ and $K_G \subset K_F$; but this is obviously not sufficient to prove that $\mathbf{Q}(\sigma)$ is not inside K_G.

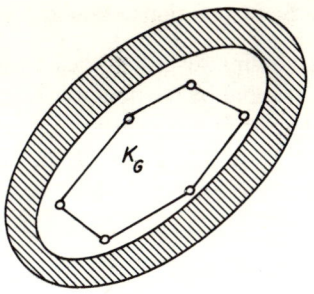

Figure V.A.5

3.3.5 Remarks

As stated in Section 3.3.2, there is no uniqueness of the function g specified in the definition of the flow rule for M. As the theorem is valid for each function g, it follows that the limit loadings are not inside

$$\bigcup_{\forall g} K_G$$

and

$$\mathbf{Q}_{\lim} \in \overline{K_F - \bigcup_{\forall g} K_G}$$

This amounts to defining the 'best' function g by a property of the envelope as indicated by Palmer [6].

3.4 Variants

The Radenkovic theorem retains all the generality of the classical limit theorems with respect to the nature of the system and the loading process. Starting from a property of the flow rule for M, the theorem is proved independently of any particular condition.

This generality leads to the possibility that the inner boundary may be trivial or present little interest. This is the case for a Coulomb material with a flow rule corresponding to $v = 0$, and for which $C_g = 0$; i.e. the G material is a liquid!

If the proof of the theorem is again considered, it is seen to require only that (26) be true globally for a limit equilibrium solution and that a field σ' verifies (24). This observation is not of use as stated, since the solution (σ, v) is assumed to be known. However, if it were possible to estimate, *a priori*, at each point of the system and for a given loading process, a likely range for the stress **σ**, then (24) and (25) would only have to be imposed for **σ** varying in this range. This would make it possible to use 'better' g functions and would improve the results of the theorem. Such procedures appear to be the aim of the works of Josselin de Jong [4, 5].

4. Friction Conditions at the Interfaces

In the case of a system involving several bodies, the friction conditions at the different interfaces are represented by particular forms of yield criteria and flow rules. They must, therefore, satisfy the principle of maximum plastic work so that the results of Section 2 are applicable. If the necessary conditions are satisfied, the theorems in Sections 3.2 and 3.3 may be used.

4.1 Various types of friction conditions

For an interface separating two materials M_1 and M_2, f_1 and f_2 are their loading functions near the interfaces and g_1, g_2 their plastic potentials, in the sense indicated in Section 3.3.

The friction conditions currently known will be examined.

4.1.1 *Smooth interface*

A smooth contact at the interface corresponds to the plastically admissible range for the stresses,

$$\sigma \leqslant 0$$
$$\tau = 0 \tag{30}$$

and to the flow rule,

$$\sigma < 0, \quad \tau = 0 \qquad [\mathbf{u}] \cdot \mathbf{n} = 0, \quad \text{any } [\mathbf{u}] \cdot \mathbf{t}$$
$$\sigma = 0, \quad \tau = 0 \qquad [\mathbf{u}] \cdot \mathbf{n} \geqslant 0, \quad \text{any } [\mathbf{u}] \cdot \mathbf{t}$$

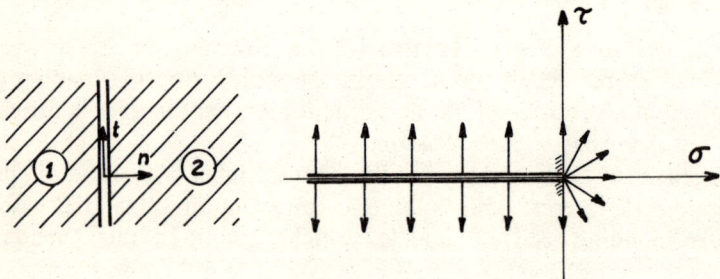

Figure V.A.6

Figure V.A.6 represents the 'loading surface' of this 'contact' and the corresponding strain rates, in this case velocity discontinuities, $[\mathbf{u}] = \mathbf{u}^2 - \mathbf{u}^1$. The principle of maximum plastic work is verified, and with the notations of Section 3.1 f is convex and $f \equiv g$, for the contact

$$f(\boldsymbol{\sigma}) = \text{Sup}\{\sigma, \tau^2\} \tag{31}$$

4.1.2 Interface with dry Coulomb friction

The plastically admissible range for the stresses is given by

$$\sigma \leq 0$$
$$|\tau| \leq \sigma \tan \phi_0 \qquad (32)$$

and the flow rule by

$$\left. \begin{array}{ll} \sigma < 0, \quad |\tau| < -\sigma \tan \phi_0 : [\mathbf{u}] = 0 \\ \sigma = 0, \quad |\tau| = -\sigma \tan \phi_0 : [\mathbf{u}] \cdot \mathbf{n} = 0, [\mathbf{u}] \cdot \mathbf{t}\tau \geq 0 \\ \sigma = 0, \tau = 0 \qquad\qquad : [\mathbf{u}] \cdot \mathbf{n} \geq 0, \quad \text{any } \mathbf{u} \cdot \mathbf{t} \end{array} \right\} \qquad (33)$$

Hence the form of Figure V.A.7.

This condition is not standard: $f \neq g$ as f corresponds to the curve of Figure V.A.7, whereas g corresponds to that of Figure V.A.4.

$$f(\boldsymbol{\sigma}) = \operatorname{Sup}\{\sigma, \tau^2 - \sigma^2 \tan^2 \phi_0\}, \quad g(\boldsymbol{\sigma}') = \operatorname{Sup}\{\sigma', \tau'^2\}$$

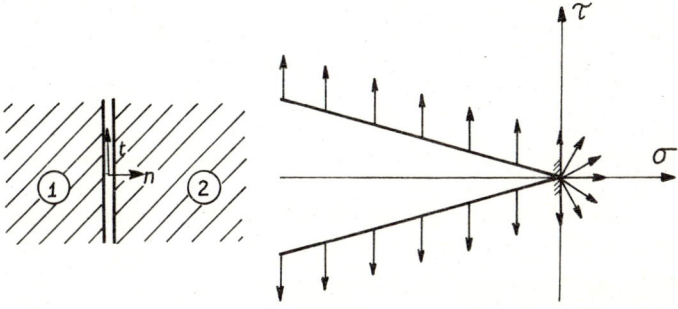

Figure V.A.7

The stress state and the velocity field at the interface are governed not only by the friction conditions at the interface itself, but also by the properties of M_1 and M_2. The plastically admissible range is therefore the intersection of (32), of $f^1 \leq 0$ and of $f^2 \leq 0$; and the velocity discontinuity is governed by (33),

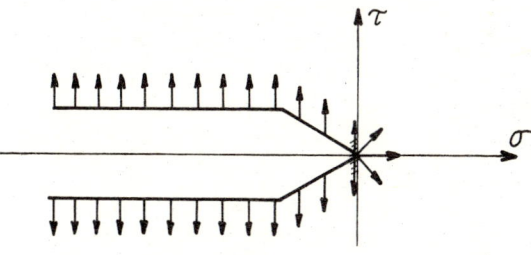

Figure V.A.8

by g^1 or by g^2, according to the stress state.[1] Figure V.A.8 represents the case in which M_1 is rigid and M_2 is a standard Tresca material.

It is assumed for the example of Figure V.A.8 that $\tan\phi \gg 1$. The condition (32) then limits the P.A. range of Figure V.A.8 to small values of $|\sigma|$. As a limiting case the scheme of Figure V.A.9 represents the so-called 'perfectly rough' interface for M_1 rigid and M_2 a standard Tresca material. A similar procedure applies for any other M_1 and M_2 materials.

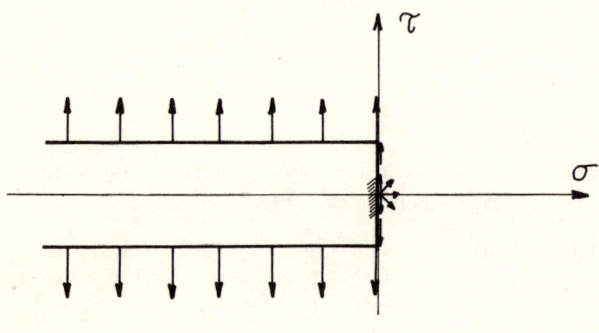

Figure V.A.9

4.1.3 Interface without relative motion

Some theorems [3] assume an ideal model in which the interface itself does not allow any relative motion. The P.A. range and the rules for the velocity discontinuities then depend only on f^1, f^2, g^1, g^2.

4.2 Application of the theorems of Section 3

4.2.1 Standard and non-standard interfaces

The study in Section 4.1 shows that only a smooth interface is standard. The 'perfectly rough' interface is not standard, regardless of the type of M_1 and M_2. (The case is similar for an interface with any dry friction condition.) The type of interface with no relative motion depends only on M_1 and M_2, and if M_1 and M_2 are standard the interface is certainly standard also.

For the 'perfectly rough' interface, or the interface with dry friction, g, regardless of M_1 and M_2, is the function corresponding to the smooth interface (see Figure V.A.6).

---

[1] Stated more precisely, f_1 and f_2 being the yield functions of M_1 and M_2, both regarding the stress tensor $\boldsymbol{\sigma}$, induce restrictions concerning the σ and τ components of the stress vector acting on the interface. Let $\phi_1 \leq 0$ and $\phi_2 \leq 0$ denote these restrictions, the plastically admissible range for (σ, τ) is therefore the intersection of (32), $\phi_1 \leq 0$ and $\phi_2 \leq 0$.

4.2.2 *System (G)*

As the convex plastic potential for the interfaces is that of a smooth interface, the application of the Radenkovic theorem is immediate. A system of standard material (G) with smooth interfaces is considered. The limit load for this system is smaller than or equal to that of the real system.[1] In particular, if the system is constituted by a standard Tresca material, then Drucker's theorem B [3] is obtained.

4.2.3 *System (F)*

The theorem of Section 3.2 leads to a maximization by considering the system made of a standard material (F) and admitting as the flow rules at the interfaces those deduced from the normality of $[\mathbf{u}]$ at the boundary of the plastically admissible range for each global interface. The limit load of this system is greater than or equal to those of the actual system.

This is the best upper bound that can be supplied by the theorem. As an alternative, as this interface flow rule is not very prepossessing, it may be preferable to consider a system composed of the same material but with perfectly rough standard interfaces (i.e. with $[\mathbf{u}]$ normal at the boundary of the P.A. range). This method is represented in Figure V.A.10 for the case of two materials, M_1, rigid, and M_2, a standard Tresca material (as in Figure V.A.10).

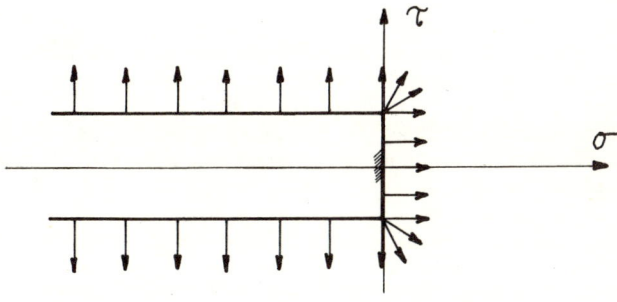

Figure V.A.10

An upper bound of the limit loads of the real system is then obtained which cannot be better than the previous value, as the convex surface of allowable loadings for this second standard system contains that of the first standard system.

This theorem is equivalent to theorem A' stated in [11]. The Drucker theorem A [3] uses the condition of an interface with no relative motion, and supplies an upper bound which cannot be better than the previous one.

---

[1] For simplicity the terms of the proportional loading case are used.

References

[1] I. F. Collins (1968) An optimum loading criterion for rigid plastic materials, *Jl. Mech. Phys. Solids*, **16**, No. 2, pp. 73–80.

[2] I. F. Collins (1969) The upper bound theorem for rigid plastic solids generalized to include Coulomb friction, *Jl. Mech. Phys. Solids*, **17**, No. 5, pp. 323–338.

[3] D. C. Drucker (1954) Coulomb friction plasticity and limit loads, *J. Appl. Mech. Trans. A.S.M.E.*, **21**, No. 1, pp. 71–74.

[4] G. de Josselin de Jong (1964) Lower bound collapse theorem and lack of normality of strain rate to yield surface for soils, *Proc. IUTAM Symp. on Rheology & Soil Mechanics*, Grenoble (Fr.), pp. 69–75.

[5] G. de Josselin de Jong (1973) A limit theorem for material with internal friction, *Proc. Symp. on the Role of Plasticity in Soil Mech.*, Cambridge (G.B.), pp. 12–21.

[6] A. C. Palmer (1966) A limit theorem for materials with non-associated flow laws, *J. Mécanique*, **5**, No. 2, pp. 217–222.

[7] D. Radenkovic (1961) Théorèmes limites pour un matériau de Coulomb à dilatation non standardisée, *C.R. Ac. Sc.*, Paris, **252**, pp. 4103–4104.

[8] D. Radenkovic (1962) Théorie des charges limites, *Séminaire de Plasticité*, Ed. J. Mandel, P.S.T. No. 116, pp. 129–142.

[9] D. Radenkovic and Nguyen Quoc Son (1972) La dualité des théorèmes limites pour une structure en matériau rigide plastique standard, *Archives of Mechanics*, **24**, No. 5–6, pp. 991–998.

[10] G. Sacchi and M. Save (1968) A note on the limit loads of non-standard materials, *Meccanica*, **3**, No. 1, pp. 43–45.

[11] J. Salençon (1969) La théorie des charges limites dans la résolution des problèmes de plasticité en déformation plane, Thesis Doct. Sc., Paris.

[12] J. Salençon (1972) Ecoulement plastique libre et analyse limite pour les matériaux standards et non standards, *Ann. I.T.B.T.P.*, No. 295–296, pp. 90–100.

[13] J. Salençon (1972) Un exemple de non validité de la théorie classique des charges limites pour un système non standard, *Proc. Int. Symp. on Foundations of Plasticity*, Warsaw 1972, North Holland Pub. Co., Amsterdam.

[14] J. Salençon (1972) Théorie des charges limites. *Sem. Plasticité et viscoplasticité*, Ed. D. Radenkovic and J. Salençon, Ediscience, Paris, 1974, pp. 207–229.

[15] J. Salençon (1972) Charge limite d'un système non-standard. Un contre exemple pour la théorie classique. *Sem. Plasticité et Viscoplasticité*, Ed. D. Radenkovic and J. Salençon, Ediscience, Paris, 1974, pp. 427–430.

[16] J. Salençon (1974a) Plasticité pour la Mécanique des Sols. C.I.S.M., Rankine session, July 1974, Udine (It.).

[17] J. Salençon (1974b) Bearing capacity of a footing on a $\phi = 0$ soil with linearly varying shear strength. *Géotechnique*, **24**, No. 3, pp. 443–446.

[18] M. Save (1961) On yield conditions in generalized stresses. *Quart. Appl. Math.*, **19**, No. 3, pp. 259–267.

B. BONNEAU'S THEOREM

The general proof of Bonneau's theorem [1] will be given for the case of a non-homogeneous material with any intrinsic curve. The existence of continuous first derivatives of the loading function with respect to x and y is assumed.

1. The Problem

A plane problem is considered.

A curve (C) drawn within a solid separates the regions (I) and (II). The normal and tangential components of the stress on (C) are given by σ and τ. These functions σ and τ along (C) are assumed, for reasons explained later, to satisfy the yield condition (see Chapter I, formula 8),

$$|\tau| = h(\sigma, x, y) \qquad (1)$$

at any point of (C), with a constant sign (i.e. $\tau = h$ or $\tau = -h$). Thus (C) is an envelope of 'marginal' surfaces of the same type.

It is intended to show that for an allowable stress-field to exist in the solid (i.e. for stress-fields to be in equilibrium in (I) and (II) while satisfying the continuity of the stress components σ and τ on (C), without violating the yield criterion at any point), it is necessary that the stresses along (C) verify the differential relation for stresses obtained in the Appendix of Chapter IV, along the characteristics corresponding to the considered type of surface. (The relations are denoted by E_α and E_β.) Thus (C) must be characteristic of the stresses.

2. Origin of the Problem—The Case of Incomplete Solutions

2.1

The problem has several origins. The first is the question of the characteristic envelopes [2], whether they be rectilineal or not. In fact, such an envelope, if touched by the arcs of characteristics in the plastic range, is the curve (C) of Section 1, along which the equation E_α (or E_β) is not verified since it is not a characteristic. Bonneau's theorem then indicates that (C) cannot be touched by the arcs of the characteristics within the plastic zone but only at its boundary, or else the envelope is not touched by the arcs of the characteristics themselves but by their prolongations.

For problems of plane deformation, the solution gives curves (C) as boundaries between the zones of type a (I) and type b (II). In this case the allowable stress-field is known in (I) (in limit equilibrium), and (C) is a characteristic or an envelope of characteristics. Bonneau's theorem gives a particularly important result in this last case, since it shows that if (C) is an envelope of characteristics within the solid, an allowable prolongation of the stress-field does not exist.

Finally, in the case of a standard material, (C) can also be a line of discontinuity of the velocity, isolated and separating two rigid zones. In this case, $\pm\tau = h(\sigma, x, y)$, the sign being determined by the sense of the velocity jump, and the stress-fields are unknown in (I) and (II). Bonneau's theorem specifies that if an allowable stress-field can be associated with this mode of deformation, it necessarily verifies equation E_α (resp. E_β) along the velocity discontinuity line.

2.2

These results have important applications for the incomplete solutions of problems of plane deformation for a standard material, as indicated below.

(1). From the viewpoint of applying the kinematic method and the use of the particular properties of the incomplete solutions, it is unnecessary for the equation E_α (or E_β) to be satisfied either at the boundary of the deformed zones in the solid or along the isolated slip lines.

An example is supplied by Ostrowska [4] which involves an envelope of characteristics touched by those in the plastic zone.

(2). If, when constructing an incomplete solution, it is hoped to go further than the simple utilization of the kinematic method and thus have an incomplete solution that can be completed to give the exact value of the limit loading, it is obvious that equation E_α (or E_β) must be satisfied at the boundary of the deformed zones in the solid and along the isolated slip lines; otherwise, the 'checking' of the rigid zones is certainly not possible.

Hill [3], led by physical reasoning, took this condition into account for the construction of incomplete solutions in a study of the indentation of a block. Assumptions were made concerning the development and localization of the plastic zones.

The theorem will now be proved.

3. Lemma

If an allowable stress-field exists in a solid, then it is continuous in crossing (C).

Proof.[1]

The allowable stress-field must satisfy the continuity of the stress components σ and τ in crossing (C).

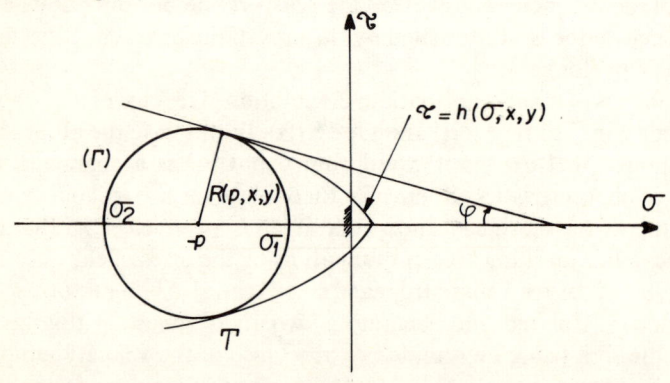

Figure V.B.1

---

[1] The principle of this proof is due to J. Mandel (private communication).

In the Mohr diagram (Figure V.B.1) the extremity T of the stress vector (σ, τ) at a point M of (C) is on the intrinsic curve of this point.

The loading function is assumed continuous with respect to x and y. It follows that at point M the intrinsic curve is identical on both sides of (C) to that on (C). Now the only Mohr circle passing through T, representative of a plastically admissible stress tensor at M, is the circle (Γ) which is tangential to the intrinsic curve of point M in T.

It may be deduced from this, since σ and τ are continuous across (C), that (Γ) represents the stress tensor at M, not only on (C) but also in the regions (I) and (II) on both sides of (C), at points M_I and M_{II}. Therefore, there is continuity of the stress tensor in crossing (C).

Corollary:

The continuity of the stress-field in crossing (C) entails continuity of the tangential derivatives of the stress along (C).[1]

4. Statement of the Differential Relations Necessarily Verified Along (C)

Consider the case of Figure V.B.2 where τ is negative.[2] The relation (1) which is satisfied along (C) is then

$$-\tau = h(\sigma, x, y)$$

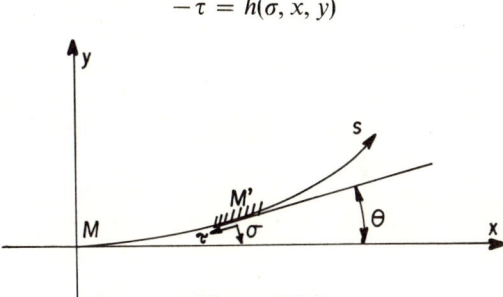

Figure V.B.2

Conditions at M are given by

$$\sigma_y = \sigma$$

$$\tau_{xy} = -\tau$$

$$\frac{\partial \sigma_x}{\partial x} = \frac{\partial \sigma_x}{\partial S}$$

and is necessarily continuous when crossing (C);

$$\frac{\partial \tau_{xy}}{\partial x} = \frac{\partial \tau_{xy}}{\partial S}$$

and, according to the lemma, is necessarily continuous when crossing (C).

[1] According to the compatibility relations of Hadamard.
[2] In the case in which (C) is an envelope of characteristics, it is an envelope of the α characteristics.

As the body forces are assumed to be continuous across (C), and since the equilibrium equations must be satisfied in (I) and (II) so that the stress-field is allowable, it follows that $\partial \tau_{xy}/\partial y$ and $\partial \sigma_y/\partial y$ are continuous across (C) at point M.

For the expression

$$g(x, y) = \tau_{xy} - h(\sigma_y, x, y) \tag{2}$$

the derivative with respect to y is given by

$$\frac{\partial g}{\partial y} = \frac{\partial}{\partial y}\{\tau_{xy} - h(\sigma_y, x, y)\} = \frac{\partial \tau_{xy}}{\partial y} - \frac{\partial h}{\partial \sigma_y}\frac{\partial \sigma_y}{\partial y} - \frac{\partial h}{\partial y} \tag{3}$$

According to the assumption of continuity for the derivatives of the loading function, and the results obtained above, $\partial g/\partial y$ is necessarily continuous when crossing (C). Now, in M,

$$g(x, y) = \tau_{xy} - h(\sigma_y, x, y) = -\tau - h(\sigma, x, y) = 0$$

According to the lemma of Section 3, g is a continuous function when crossing (C). Also, if the loading function of the material is denoted by f, so that the stress-field is allowable in (I) and (II), it is necessary in both regions to have

$$f(\boldsymbol{\sigma}, x, y) \leqslant 0$$

Now

$$f(\boldsymbol{\sigma}, x, y) \leqslant 0 \Rightarrow \tau_{xy} - h(\sigma_y, x, y) \leqslant 0.$$

Therefore, the only possibility for

$$\frac{\partial}{\partial y}\{(\tau_{xy} - h(\sigma_y, x, y))\}$$

to be continuous at M when crossing (C) is for it to be zero:[1]

$$\frac{\partial}{\partial y}\{(\tau_{xy} - h(\sigma_y, x, y))\} = 0 \tag{4}$$

According to Figure V.B.1,

$$\frac{\partial h}{\partial \sigma} = -\tan \phi$$

Hence, by substitution in (3),

$$\frac{\partial \tau_{xy}}{\partial y} + \tan \phi \frac{\partial \sigma_y}{\partial y} - \frac{\partial h}{\partial y} = 0 \quad \text{at} \quad M \tag{5}$$

It is convenient to express $\partial/\partial y[h(\sigma, x, y)]$ as a function of $\partial/\partial y[R(p, x, y)]$, introduced in the Appendix A of Chapter IV:

$$\frac{\partial}{\partial y}h(\sigma, x, y) = \frac{1}{\cos \phi}\frac{\partial}{\partial y}R(p, x, y) \tag{6}$$

---

[1] Thus the stress-field is not only at the yield boundary on (C) and continuous across it, but also, on both sides near (C), it is necessarily tangential to a field at the yield boundary.

Then (5) can be transformed by taking (6) into account and introducing the derivatives with respect to x of the stresses by means of the equations of equilibrium. The components of the body forces in M following Mx and My are given by X and Y. Hence,

$$\frac{\partial \sigma_x}{\partial x} + \tan \phi \frac{\partial \tau_{xy}}{\partial x} + \rho(X + Y \tan \phi) + \frac{1}{\cos \phi} \frac{\partial R}{\partial y} = 0 \qquad (7)$$

Let M' be a point on (C) in the vicinity of M, set $p = -(\sigma_1 + \sigma_2)/2$, and denote by θ the angle of the direction of greatest tension at M' with Mx. Then, at M',

$$\begin{cases} \sigma_x = p + R \cos 2\theta \\ \tau_{xy} = R \sin 2\theta \end{cases}$$

and hence, since

$$\frac{\partial R}{\partial p} = \sin \phi$$

$$\begin{cases} \dfrac{\partial \sigma_x}{\partial x} = -\dfrac{\partial p}{\partial x}(1 - \sin \phi \cos 2\theta) - 2R \sin 2\theta \dfrac{\partial \theta}{\partial x} + \cos 2\theta \dfrac{\partial R}{\partial x} \\ \dfrac{\partial \tau_{xy}}{\partial x} = \dfrac{\partial p}{\partial x} \sin \phi \sin 2\theta + 2R \cos 2\theta \dfrac{\partial \theta}{\partial x} + \sin 2\theta \dfrac{\partial R}{\partial x} \end{cases}$$

and at a point M where $\theta = \pi/4 + \phi/2$

$$\begin{cases} \dfrac{\partial \sigma_x}{\partial x} = -\dfrac{\partial p}{\partial x}(1 + \sin^2 \phi) - 2R \cos \phi \dfrac{\partial \theta}{\partial x} - \sin \phi \dfrac{\partial R}{\partial x} \\ \dfrac{\partial \tau_{xy}}{\partial x} = \dfrac{\partial p}{\partial x} \sin \phi \cos \phi - 2R \sin \phi \dfrac{\partial \theta}{\partial x} + \cos \phi \dfrac{\partial R}{\partial x} \end{cases}$$

Thus (7) becomes

$$-\frac{\partial p}{\partial x} - 2 \frac{R}{\cos \phi} \frac{\partial \theta}{\partial x} + \rho(X + Y \tan \phi) + \frac{1}{\cos \phi} \frac{\partial R}{\partial y} = 0 \qquad (8)$$

$(X + Y \tan \phi) = F^\alpha$ is the component of the body force for the unit basis e_α, e_β, where e_α is tangential at M to Mx and e_β is defined by

$$(e_\alpha, e_\beta) = \frac{\pi}{2} + \phi \qquad \frac{1}{\cos \phi} \frac{\partial R}{\partial y} = \frac{\partial R}{\partial X_\beta}$$

using the notation of the appendix of Chapter IV. Thus, the following equation is obtained.

$$dp + \frac{2R}{\cos \phi} d\theta - \left(\rho F^\alpha + \frac{\partial R}{\partial x_\beta} \right) dx^\alpha = 0 \text{ along } (C)$$

which is simply the equation E_α. (See Appendix A of Chapter IV.)

It is obvious that if the case of $+\tau = h(\sigma, x, y)$ had been considered along (C), the equation E_β would have been obtained.

155

5. Remarks

5.1

The reasoning given in Section 4 indicates the significance of the equation E_α (or E_β). *Along (C) the extremity of the stress vector acting on (C) is on the intrinsic curve and the equations of equilibrium are satisfied on both sides of (C), with the yield criterion not being violated.*

This reasoning can be taken as a basis for a different introduction of the stress characteristics in the plastic zone. At any point of the plastic zone, the two 'marginal' faces on which $|\tau| = h(\sigma, x, y)$ (faces corresponding to the points of contact of the Mohr circle and the intrinsic curve) are to be considered. If (C) is a curved envelope of the faces of a family, then one of the equations E_α (or E_β) must be satisfied along (C). This equation is a differential relation between the unknown functions, and therefore (C) is a characteristic of the stress equations.

5.2

The flow rule of the material does not appear in the above proofs. The considerations concern only the stresses, and express the possibility of the existence of an allowable stress-field in the indicated conditions. It follows that the result obtained is valid for non-standard materials, and also for elasto-plasticity.

References

[1] M. Bonneau (1947) Equilibre limite et rupture des milieux continus. *Ann. Ponts et Chaussées*, Sept–Oct. 1947, pp. 609–653, Nov–Dec. 1947, pp. 769–801.
[2] J. R. Booker (1970) A property of limiting lines for a perfectly plastic material. Univ. Sydney Civ. Eng. Lab., Res. Rept. No. R 134, March 1970.
[3] R. Hill (1950) A theoretical investigation of the effect of specimen size in the measurement of hardness. *Phil. Mag.*, **41**, pp. 745–753.
[4] J. Ostrowska (1967) Solution of indentation problem with envelope of slip lines. *Bul. Ac. Pol. Sc. série Sc. Tech.*, **15**, No. 10, pp. 603–611.

Index

The numbers refer to the chapter (Roman figures), and to the paragraph within the chapter (Arabic figures)

A

Active pressure—IV.24
Anisotropy—I.2
Associated (—flow rule)—I.7
Association (Theorem of —) V App. A
Axial symmetry (uncontained plastic flow in —) IV App. B

B

BONNEAU (-'s theorem) V App. B
Boundary (elastic —)—II.3
Boundary (yield—)—II.3
Brittle (material, fracture) I.2

C

CAQUOT—I.2
Cavities (spherical and cylindrical—)—II.5
Characteristic lines—IV.6
Characteristics (Method of—) IV.8
Characteristics (Relations along the stress —) IV.7
Characteristics (Relations along the velocity—) IV.13
Coaxiality—I.10
Complete solution—V.7
Consistent boundary conditions—III.5
Constitutive law—I.5
Contained (—plastic flow)—II.2
Convexity (of the yield function)—I.2
Corresponding states (theorem of the—)—IV.23
COULOMB (-'s yield criterion)—I.2
COULOMB ('s wedge)—V.9
Criterion (yield—)—I.1
Critical (load)—II.4
Curve (intrinsic—)—I.2

D

Deviator (stress—)—I.2
Discontinuity (stress—)—IV.16
Discontinuity (velocity—)—IV.15
Dislocations—I.4
Dissipation—V.2
Domain of elasticity—II.3
DRUCKER (-'s postulate)—I.7
DRUCKER and PRAGER (-'s yield criterion)—I.2

E

Elasto-plasticity—II
Envelope (of characteristics)—V App. B

F

FELLENIUS (-'s method)—V.9
Finite elements (—for limit analysis)—V.9
Flow (—rule)—I.5
Fracture—I.2

G

GEIRINGER (-'s equations)—IV.13
GVOZDEV (-'s theorem)—V.3

H

HAAR-KARMAN (—hypothesis)—IV App. B
Hardness (—test)—IV.20
HENCKY (-'s net)—IV.10
HENCKY (-'s relation)—IV.9
HENCKY (-'s theorem)—IV.10
HILL (-'s principle)—I.6
Homogeneous (—stress-field)—IV.10

I

Interface (conditions at the—)—V App. A.4
Irreversibility—I.1
Isotropy—I.2

K

Kinematic approach—V.4
Kinematic method—V.4
Kinematic solution—V.7
KÖTTER (-'s equations)—IV.7

L

Limit analysis—V
Limit equilibrium—III.2
Limit load—II.3
Limit (elastic—of a system)—II.3
Loading process—II.3
Loading (—surface)—I.1

M

MANDEL—IV.4
MANDEL (-'s equations)—IV.7
MASSAU (-'s method)—IV.8
Minimum (—principle for the strain rates)—V.6
Minimum (—principle for the stresses)—V.6

N

Non-homogeneous (Plane strain for—material)—IV. App. A
Non-standard (—system)—V App. A.3
Normality—I.6
N_c, N_q, N_γ—IV.23

P

Parameters (loading—)—III.5
Passive pressure—IV.24
Path (loading—)—II.2
Perfect plasticity—I.1
Permanent (—deformation)—I.1
Permissible (—stress-field)—V.2
Permissible (—velocity field)—V.2
Permissible (—loading)—V.3
Plane (—strain)—IV
Plastic (—deformation)—I.3
Plastic (—potential)—I.6
PRANDTL and REUSS (—constitutive equation)—II.1

R

RANKINE (—equilibrium)—IV.24
Rigid-plastic material—III.1
Rule (flow—)—I.5

S

SAINT-VENANT (-'s hypothesis)—I.5
Semi-homogeneous (—stress field)—IV.10
SOKOLOVSKI—IV.24
Standard material—I.8
Static approach—V.3
Static method—V.3
Static solution—V.7
Strain-hardening—I.1
Strain-hardening (of a system)—II.3
Strain softening—II.2
Superposition (Method of—)—IV.23

T

TERZAGHI (-'s formula)—IV.23
TRESCA (-'s yield criterion)—I.2

U

Uncontained (—plastic flow)—II.2
Uniqueness (—of the stress field)—III.6
Uniqueness (—of the limit loadings)—V.3
Unloading—I.5

V

Visco-plasticity—I.5

Y

Yield (—boundary)—II.3
Yield (—criterion)—I.1
Yield (—function)—I.1
Yield (—limit)—I.1

TA
710
S2413

OCT 5 1978